SMITHSONIAN

METEOROLOGICAL OBSERVATIONS

FOR THE YEAR

1855.

[PRINTED FOR EXAMINATION BY THE OBSERVERS.]

WASHINGTON, D. C.

1857.

This edition of abstracts of the Meteorological Observations is printed to ascertain the form best adapted to the purpose, and is referred to the Observers for critical examination.

JOSEPH HENRY,

Secretary Smithsonian Institution.

WASHINGTON, *November*, 1857.

NAME OF STATION.	North Latitude.	West Longitude from Greenwich	Height above the Sea.	Months observed.	NAME OF OBSERVER.
	° ′	° ′	FEET.	NUM.	
Aiken, S. C.	33 32	81 34	565	12	H. W. Ravenel.
Alexandria, Va	38 48	77 1	56	12	Benjamin Hallowell.
All Saints, S. C. (1)	33 40	79 17	20	12	Rev. A. Glennie.
Amherst, Mass	42 22	72 34	267	12	Prof. E. S. Snell.
Angelica, N. Y	42 15	78 1	1500	9	Dr. E. M. Alba.
Ann Arbor, Mich. (2)	42 15	83 30	....	12	L. Woodruff.
Ann Arbor, Mich. (3)	42 16	83 44	891	12	Prof. A. Winchell.
Annapolis, Md	38 59	76 30	34	2	Dr. A. Zumbrock.
Arcola, Ohio	41 55	81 6	....	2	Ardelia Cunningham.
Ashland, Va. (4)	38 38	81 57	....	9	Samuel Couch.
Athens, Ill	39 52	89 56	....	12	Joel Hall.
Auburn, Ala	32 37	85 36	....	11	Prof. J. Darby.
Augusta, Ga	38 28	81 54	....	6	William Haines.
Augusta, Ill	40 12	90 45	....	12	Dr. S. B. Mead.
Austin, Texas	30 20	97 46	....	12	Dr. Sam. K. Jennings, Jr.
Baldwinsville, N. Y	43 4	76 41	....	12	John Bowman.
Baldwin Institute, Ohio (5)	....	....	....	1	Prof. G. M. Barber.
Ballardsville, Ky	38 24	85 31	....	3	Dr. John Swain.
Battle Creek, Mich	42 20	85 1	....	12	William N. Campbell.
Bedford, Pa	40 1	78 30	....	10	Samuel Brown.
Beloit, Wis	42 30	89 4	750	12	Prof. William Porter.
Bellefontaine, Wis	40 20	83 46	....	1	Joseph Shaw.
Beverly, N. Y. (6)	41 23	72 12	180	12	Thomas B. Arden.
Blackwell's Island, N. Y	41 14	74 1	....	1	Dr. William W. Sanger.
Bladensburg, Md	38 57	76 58	....	11	B. O. Lowndes.
Bloomfield, N. J	40 49	74 11	120	12	Robert L. Cooke.
Boston, Mass	42 22	71 3	....	10	
Bowling Green, Ky.	37	86 25	....	2	J. C. Herrick.
Brandon, Vt	43 45	73 8	....	12	David Buckland.
Burlington, Vt. (7)	44 29	73 11	346	12	Prof. Zadock Thompson.
Burlington, N. J	40	75 12	26	12	Rev. Adolph Frost.
Camden, S. C	34 17	80 33	275	12	Dr. J. A. Young.
Camden, Ark	33 32	92 48	....	1	J. J. McElrath.
Canton, N. Y	44 38	76 15	304	12	E. W. Johnson.
Carlisle, Pa	40 12	77 11	500	1	Prof. W. C. Wilson
Carmel, Me	44 47	69	175	12	John J. Bell.
Castleton, Vt	43 34	73 9	....	5	D. Underwood.
Cedar Keys, Fla. (8)	29 8	83 3	....	12	Augustus Steele.
Ceresco, Wis	43 50	88 57	917	10	Miss Mary E. Baker.
Chapel Hill, N. C. (9)	35 54	79 17	....	12	Prof. James Phillips.
Charleston, S. C	32 46	79 57	....	2	
Cheviot, Ohio	39 7	84 34	....	3	Ebenezer Hannaford.
Chestertown, Md	39 14	76 2	....	6	James Alfred Pearce, Jr.
Cincinnati, Ohio	39 6	84 15	....	10	Boston Telegraph.
Cincinnati, Ohio	39 6	84 25	....	1	John Lee.
Cleveland, Ohio	41 35	81 44	....	8	G. A. Hyde.
Columbus, Miss	33 30	88 29	300	12	James S. Lull.
Concord, N. H	43 12	71 29	374	12	Dr. William Prescott.
Cooper, Mich	42 40	85 31	....	11	Mrs. Octavia L. Walker.
Craftsbury, Vt	44 40	72 29	1100	9	James A. Paddock.
Crichton's Store, Va	41 20	77 50	....	11	Lieut. R. F. Astrop.
Danville, Ky	37 40	84 31	950	9	O. Beatty.
Dayton, Ohio	39 44	84 10	....	1	Dr. James C. Fisher.
D. & Dumb Inst. N. Y. (10)	40 43	74 5	159	12	Prof. O. W. Morris.

(1) Waccaman.
(2) 3½ miles E.S.E. from Ann Arbor.
(3) University of Michigan.
(4) Near Buffalo.
(5) Berea.
(6) Phillipstown.
(7) University of Vermont.
(8) Atsena.
(9) University of North Carolina.
(10) New York City.

NAME OF STATION.	North Latitude.	West Longitude from Greenwich	Height above the Sea.	Months observed.	NAME OF OBSERVER.
	° ′	° ′	FEET.	NUM.	
Delaware College, (1)	39 38	75 47	120	4	Prof. W. A. Crawford.
Detroit, Mich	42 24	83	620	12	Rev. George Duffield.
Dubuque, Iowa	42 29	90 50	680	12	Dr. Asa Horr.
Easton, Pa. (2)	40 43	75 16	....	12	Selden J. Coffin.
Exeter, N. H	42 58	70 55	....	12	Rev. L. W. Leonard.
Flushing, N. Y	40 46	73 52	....	6	
Fort Madison, Iowa	40 37	91 28	....	12	Daniel McCready.
Flint, Mich	42 58	83 39	....	12	Drs. D. Clark and M. Miles.
Frederick City, Md	39 24	77 18	....	12	Henry E. Hanchew.
Friendship, Tenn	35 50	89 25	....	4	Dr. R. T. Turner.
Fryeburg, Me	44 3	71	....	12	George B. Barrows.
Gallipolis, Ohio	39	82 1	520	11	G. W. Livesay.
Gardiner, Me	44 11	69 46	75	8	R. H. Gardiner.
Garlandsville, Miss	32 23	89 20	....	5	Rev. E. S. Robinson.
Germantown, Ohio	39 30	84 20	720	12	L. Groneweg.
Gettysburg, Pa	39 51	77 15	....	12	Prof. M. Jacobs.
Glenwood, Tenn. (3)	36 28	87 13	481	12	William M. Stewart.
Gouverneur, N. Y	44 25	75 35	....	12	P. O. Williams.
Grand Rapids, Mich	43	86	743	12	Alfred E. Currier.
Granville, Ohio	40 4	82 34	....	12	Prof. S. N. Sanford.
Green Springs, Ala. (4)	32 50	87 46	....	10	Prof. H. Tutwiler.
Great Falls, N. H	43 18	70 52	250	6	H. E. Sawyer.
Hamilton, Canada	43 15	79 56	....	6	Dr. W. Craigie.
Hannibal, Mo. (5)	39 44	91 23	....	5	O. H. Lear.
Harrisburg, Pa	40 16	76 50	....	12	Dr. John Heisely.
Hillsborough, Ohio	39 13	83 30	1000	12	Prof. J. McD. Matthews.
Hiram, Ohio	41 20	81 15	765	2	S. L. Hillier.
Jackson, Miss	32 23	90 8	....	3	Hatch & Co.
Jackson, Ohio	39 7	82 30	700	8	George L. Crookham.
Jackson, Ohio	39 7	82 30	....	8	M. Gilmor.
Jackson, Ohio	39 7	82 30	....	2	S. B. Wood.
Jacksonville, Fla	30 15	82	14	12	Dr. A. G. Baldwin.
Janesville, Wis	42 42	89 91	768	3	J. F. Willard.
Keokuk, Iowa	40 25	91 21	....	1	Miss Ida E. Ball.
Key West, Fla. (6)	24 33	81 48	4	12	W. C. Dennis.
Knox Hill, Fla. (7)	30 30	86 1	148	12	John Newton.
Knoxville, Tenn. (8)	35 59	83 54	1000	6	Prof. T. L. Griswold.
La Grange, Ga	33 2	84 55	....	1	
Lac qui Parle, Min. (9)	45	95 30	....	8	Rev. S. R. Riggs.
Lebanon, Tenn	36 15	86 15	....	1	Prof. Benjamin C. Jillson.
Lewisburg, Va	37 49	80 28	....	12	Dr. Thomas Patton.
Lima, Pa. (10)	39 55	75 25	196	12	Joseph Edwards.
Lodi, N. Y	42 37	76 53	....	12	John Lefferts.
Londonderry, N. H	42 53	71 20	....	10	R. C. Mack.
Lowville, N. Y	43 45	75 38	....	7	J. Carroll House.
Madison, Ind	....	....	....	6	C. Barnes.
Madison, Ohio	41 55	81 06	725	1	Miss Ardelia Cunningham.
Madison, Wis	43 5	89 25	892	1	Prof. J. W. Sterling.
Madrid, N. Y	44 43	75 33	280	3	E. A. Dayton.
Manchester, Ill	39 33	90 35	723	12	John Grant.
Manchester, N. H	42 59	71 28	300	12	Hon. Samuel N. Bell.
Marietta, Ohio	39 25	81 29	....	5	Prof. J. W. Andrews.
Meadville, Pa	41 39	80 11	....	8	J. F. Thickstun.
Mendon, Mass	42 6	71 34	....	12	Dr. John G. Metcalf.

(1) Newark, Del.
(2) La Fayette College.
(3) Near Clarksville.
(4) Havana.
(5) Dry Ridge.
(6) Salt Ponds, $2\frac{1}{2}$ miles E.N.E. from Key West.
(7) Orange Hill.
(8) East Tennessee University.
(9) Hazelwood.
(10) Chromedale, near Lima.

NAME OF STATION.	North Latitude.	West Longitude from Greenwich	Height above the Sea.	Months observed.	NAME OF OBSERVER.
	° ′	° ′	FEET.	NUM.	
Millersburg, Ky.	38 40	84 27	804	9	Rev. G. S. Savage.
Milton, Ind.	39 47	85 2	800	12	Dr. V. Kersey.
Milwaukie, Wis.	43 4	87 57	593	12	Carl Winkler.
Milwaukie, Wis. (1)	43 4	87 59	658	2	F. C. Pomeroy.
Monroe, Mich.	41 56	83 23	590	12	H. Whelpley.
Monroe, Mich.	41 58	83 24	....	2	Capt. A. D. Perkins.
Monroeville, Ala.	31 33	87 25	....	7	S. J. Cumming.
Montcalm, Va.	38 5	78 21	450	1	Charles J. Meriwether.
Montreal, Canada	45 31	73 33	118	5	Dr. A. Hall.
Morrisville, Pa.	40 12	74 53	30	12	Ebenezer Hance.
Moss Grove, Pa. (2)	41 40	79 51	....	12	Francis Schreiner.
Mount Harris, Me.	44 40	69 8½	1200	....	L. McLean Titton.
Mount Vernon, Ohio	40 25	82 31	....	4	F. A. Benton.
Muscatine, Iowa	41 26	91 5	586	12	T. S. Parvin.
Nantucket, Mass.	41 17	70 6	30	12	Hon. William Mitchell.
Nazareth, Pa.	40 43	75 21	....	1	H. A. Brinkenstein.
Newark, Ohio	40 6	82 28	....	5	Lewis M. Dayton.
Newark, N. J.	40 45	74 10	....	12	W. A. Whitehead.
New Bedford, Mass.	41 39	70 56	90	12	Samuel Rodman.
Newburyport, Mass.	42 49	70 53	....	12	Dr. H. C. Perkins.
New Harmony, Ind.	38 8	87 9	320	12	John Chappellsmith.
New Lisbon, Ohio	40 45	80 46	....	5	
New London, Ct.	41 22	72 9	90	10	Rev. T. Edwards, D. D.
New Orleans, La.	29 57	90	....	12	Dr. E. H. Barton.
New Wied, Texas	29 42	98 15	....	12	Prof. L. C. Ervendberg.
Norristown, Pa.	40 8	75 19	153	12	Rev. Grier Ralston.
North Attleboro, Mass.	41 52	71 23	175	12	Henry Rice.
Norwalk, Ohio	41 13	82 43	....	1	Gustavus A. Hyde.
Oberlin, Ohio	41 20	82 15	800	12	Prof. J. H. Fairchild.
Ogdensburg, N. Y.	44 43	75 26	....	11	W. E. Guest.
Oldtown, Me. (3)	44 48	68 45	127	8	Rev. Samuel H. Merrill.
Ottawa, Ill.	41 20	88 47	....	12	Dr. J. O. Harris.
Oswego, N. Y.	43 25	76 35	....	11	Capt. S. Malcolm.
Oxford, Miss. (4)	34 20	89 25	300	12	Prof. L. Harper.
Pella, Iowa	41 30	92 55	....	12	E. H. A. Scheeper.
Pen Yan, N. Y.	42 42	77 7	740	12	Dr. H. Sartwell.
Perry, Me.	45	67 6	100	12	William D. Dana.
Perrysburg, Ohio	41 39	83 40	....	12	F. Hollenbeck.
Philadelphia, Pa.	39 57	75 11	53	12	Prof. Jas. A. Kirkpatrick.
Pittsburg, Pa.:					
Oakland Station	40 32	80 2	850	12	W. W. Wilson.
Marine Hospital	....	....	....	12	W. Martin and J. Hastings.
Platteville, Wis.	42 45	90	....	12	Dr. J. L. Pickard.
Pocopson, Pa.	39 54	75 37	218	12	Fenelon Darlington.
Pomfret, Conn.	41 52½	72	596	12	Rev. D. Hunt.
Port Gibson, Miss.	31 50	91	....	1	Prof. J. B. Elliot.
Portland, Me.	43 37	70 19	....	3	Henry Willis.
Pottsville, Pa.	40 41	76 24	....	12	Dr. A. Heger.
Poultney, Iowa, (5)	42 40	91 21	....	12	Dr. Benjamin F. Odell.
Princeton, Mass.	42 28	71 53	1113	12	Hon. John Brooks.
Providence, R. I. (6)	41 49	71 25	120	10	Prof. A. Caswell.
Quasqueton, Iowa	42 23	91 43	888	10	Dr. Edwin C. Bidwell.
Randolph, Pa.	41 28	80 10	1720	7	O. T. Hobbs.

(1) Two miles north of the centre of the city.
(2) Kingsleys.
(3) Blue Hill.
(4) University of Mississippi.
(5) Plum Spring.
(6) Brown University.

NAME OF STATION.	North Latitude.	West Longitude from Greenwich	Height above the Sea.	Months observed.	NAME OF OBSERVER.
	° ′	° ′	FEET.	NUM.	
Red River Settlement, Hudson's Bay Territory......	58 6	97	853	7	Donald Gunn.
Red Wing, Min...............	44 34	92 30	....	2	Rev. Jabez Brooks.
Richmond, Mass. (1).......	42 24	73 20	....	11	William Bacon.
Richmond, Ohio............	40 20	81 1	....	10	Jacob A. Desellem.
Sacramento, Cal............	38 35	121 40	....	3	Dr. F. W. Hatch.
Sacramento, Cal............	38 35	121 40	30	3	Dr. T. M. Logan.
Sag Harbor, N. Y...........	41	72 20	40	11	Ephraim A. Byram.
St. Augustine, Fla.........	29 54	81 26	....	1	Dr. P. D. Mauran.
St. James, Mich.............	45 44	85 27	598	12	James J. Strang.
St. Johnsbury, Vt..........	44 30	74 1	....	6	James K. Colby.
St. Joseph, Min.............	48 55	98	....	2	O. A. Kellum.
St. Louis, Mo................	38 37	90 15	432	12	Dr. George Engelman.
St. Martin, Canada East...	45 32	73 s6	118	12	Dr. Charles Smallwood.
Salmon Falls, N. H.........	43 30	71	....	1	George B. Sawyer.
San Francisco, Cal.........	38	122	....	1	Dr. H. Gibbons.
Saugatuck, Mich............	40 30	85 50	....	11	L. H. Streng.
Savannah, Ohio............	41 12	82 31	....	9	Dr. John Ingram.
Savannah, Ga...............	32 5	81 7	42	12	Dr. John F. Posey.
Saybrook, Conn. (2)........	41 16	72 20	10	11	James Rankin.
Schellman Hall, Md. (3)...	39 23	76 57	700	12	Miss Harriet M. Baer.
Seneca Coll. Inst. N. Y. (4)	42 41	76 52	....	2	J. W. Chickering, Jr.
Smithfield, Va..............	37	76 40	....	12	Dr. J. R. Purdie.
Smithsonian Institution...	38 53	77 1	....	12	H. Diebitsch.
Smithville, N. Y............	44	76 1	....	12	J. E. Breed.
Southwick, Mass............	42 2	72 10	265	12	A. Holcomb.
South Thomaston, Me......	44 6	69 12	....	1	Joshua Bartlett.'
Sparta, Georgia.............	33 17	83 9	550	12	Dr. & Mrs. E. M. Pendleton.
Spencertown, N. Y.........	43 19	73 41	800	12	A. W. Morehouse.
Springdale, Ky..............	38 7	85 24	570	11	Mrs. Lawrence Young.
Springfield, Mass...........	42 6	72 35	199	12	Lucius C. Allin.
Stratford, N. H..............	44 40	71 34	....	5	William B. G. Brown.
Steuben, Me..................	44 44	67 50	50	12	J. D. Parker.
Superior, Wis...............	46 38½	92 3½	658	7	William H. Newton.
Taunton, Mass...............	41 49	71 9	....	1	Albert Schlegel.
The Rock, Ga...............	32 54	84 24	833	12	Dr. James Anderson.
Thornbury, N. C. (5).......	36 20	77 21	....	4	Daniel Morelle.
Tuscaloosa, Ala. (6)........	33 12	87 28	245	3	George Benagh.
Unionville, Ohio............	41 50	81	....	7	Miss Ardelia Cunningham.
Upper Alton, Ill.............	39	89 36	....	12	Dr. John James.
Urbana, Ohio................	40 6	83 43	1015	9	Prof. M. G. Williams.
Wampsville, N. Y...........	43 4	75 50	500	12	Dr. Stillman Spooner.
Warrington, Fla. (7)........	30 21	87 16	12	12	John Pearson.
Watertown, N. Y............	43 56	74 55	....	1	Dr. P. O. Williams.
Westfield, Mass.............	42 6	72 44	....	12	Rev. Emerson Davis, D. D.
Wess Haverford, Pa........	40	75 21	....	12	Dr. Paul Swift.
Whitemarsh Island, Ga....	32	81	18	12	R. T. Gibson.
Williamstown, Mass. (8)...	42 43	71 13	....	12	James Orton.
Windham, Me...............	43 49	70 17	....	2	S. A. Eveleth.
Winchester, Va..............	39 15	78 10	....	9	John W. Marvin.
Wood's Hole, Mass. (9)....	41 40	70 17	25	3	R. R. Gifford.
Worcester, Mass............	42 16	71 48	536	12	Edward F. Smith.
Woodward H. School, (10)	39 6	84 25	....	5	F. W. Hurtt.
Zanesville, Ohio............	39 58	82 29	700	11	John G. F. Holston.

(1) Elmwood.
(2) Lynde Point.
(3) Sykesville.
(4) Ovid.
(5) Near Jackson.
(6) University of Alabama.
(7) Navy Yard at Pensacola.
(8) Williams College.
(9) Falmouth.
(10) Cincinnati, Ohio.

NAME OF STATION.	Mean.			Maxima.			Minima.		
	7 A. M.	2 P. M.	9 P. M.	7 A. M.	2 P. M	9 P. M.	7 A. M.	2 P. M.	9 P. M.
Alexandria, Va.	30.11	30.07	30.10	30.77	30.80	30.76	29.39	29.35	29.33
All Saints, S. C.	30.09	30.05	30.06	30.52	30.56	30.44	29.47	29.45	29.44
Amherst, Mass.	29.86	29.80	29.84	30.66	30.60	30.57	28.88	29.04	28.76
Ann Arbor, Mich.	29.03	29.00	29.00	29.81	29.70	29.76	28.37	28.29	28.09
Auburn, Ala.	29.42	29.38	29.41	29.83	29.80	29.80	28.89	28.76	28.98
Ballardsville, Ky.	29.08	29.07	29.08	29.74	29.63	29.74	28.41	28.21	28.50
Bedford, Pa.	29.15	29.14	29.15	29.89	29.85	29.80	28.50	28.51	28.50
Beloit, Wis.	29.18	29.13	29.16	29.79	29.75	29.81	28.82	28.64	28.76
Bloomfield, N. J.	29.92	29.89	29.93	30.63	30.66	30.66	29.00	28.98	28.95
Burlington, Vt.	29.68	29.65	29.68	30.39	30.48	30.49	28.65	28.75	28.83
Burlington, N. J. (2)	30.10	30.09	30.12	30.70	30.78	30.76	29.30	29.44	29.27
Camden, S. C.	29.98	29.94	29.95	30.43	30.47	30.49	29.34	29.33	29.23
Canton, N. Y.	29.44	29.39	29.46	30.24	30.27	30.23	28.56	28.68	28.72
Carmel, Me.	29.67	29.66	29.66	30.65	30.61	30.53	28.83	28.90	28.91
Cedar Keys, Fla.	29.99	29.97	29.96	30.30	30.26	30.25	29.65	29.52	29.56
Ceresco, Wis.	29.05	29.03	29.03	29.57	29.62	29.66	28.71	28.65	28.55
Chapel Hill, N. C.	29.55	29.51	29.51	30.03	30.07	29.96	28.93	28.97	28.88
Charleston, S. C. (1)	......	......	......	......	......	......	......	......	......
Concord, N. H.	29.73	29.68	29.71	30.60	30.54	30.51	28.68	28.77	28.61
Deaf & Dumb Inst., N. Y.	30.00	29.97	30.00	30.69	30.72	30.67	29.11	29.06	29.05
Detroit, Mich.	29.49	29.47	29.47	30.30	30.21	30.23	28.81	28.80	28.54
Dubuque, Iowa	29.30	29.24	29.23	29.90	29.92	29.93	28.94	28.83	28.96
Frederick, Md.	29.78	29.74	29.76	30.46	30.45	30.44	29.08	29.08	29.01
Gardiner, Me.	29.92	29.92	29.92	30.75	30.73	30.66	28.93	29.15	29.12
Germantown, Ohio	29.25	29.20	29.23	29.97	29.83	29.96	28.62	28.25	28.54
Gettysburg, Pa.	29.51	29.47	29.48	30.20	30.18	30.13	28.78	28.80	28.79
Glenwood, Tenn.	29.62	29.57	29.60	30.13	30.18	30.23	28.77	28.94	28.96
Granville, Ohio	28.97	28.95	28.94	29.72	29.57	29.69	28.43	28.05	28.02
Great Falls, N. H.	29.60	29.59	29.61	30.44	30.42	30.38	28.67	28.68	28.70
Hamilton, Canada (3)	29.65	......	29.65	30.50	......	30.40	28.83	......	28.80
Harrisburg, Pa.	29.74	29.74	29.77	30.40	30.34	30.38	29.05	29.09	29.08
Hillsborough, Ohio	28.87	28.84	28.86	29.57	29.45	29.56	28.28	27.92	28.14
Jacksonville, Fla.	30.20	30.15	30.17	30.59	30.54	30.60	29.75	29.58	29.66
Key West, Fla.	29.80	29.84	......	30.01	30.05	......	29.65	29.65	......
Knox Hill, Fla. (4)	29.82	29.77	29.81	30.08	30.03	30.12	29.41	29.31	29.54
Knoxville, Tenn.	29.09	29.05	29.08	29.55	29.53	29.57	28.46	28.24	28.58
La Grange, Ga. (7)	29.11	29.10	29.08	29.52	29.51	29.48	28.56	28.49	28.52
Lima, Pa.	29.94	29.90	29.94	30.62	30.62	30.62	29.10	29.04	29.13
Madison, Ia.	29.49	29.45	29.47	30.15	30.10	30.18	28.71	28.52	28.95
Manchester, N. H.	30.10	30.09	30.08	30.91	30.86	30.83	29.17	29.22	29.14
Marietta, Ohio	29.23	29.19	29.22	29.94	29.77	28.88	28.72	28.41	28.46
Milton, Ia	29.02	29.00	29.03	29.64	29.60	29.67	28.40	28.07	28.26
Milwaukie, Wis.	29.25	29.22	29.20	29.92	29.95	29.97	28.81	28.68	28.72
Montcalm, Va. (5)	29.39	29.36	29.39	29.59	29.65	29.65	29.09	28.78	29.24
Monroe, Mich. (2)	28.88	28.90	28.88	29.60	29.65	29.50	28.20	28.30	28.00
Morrisville, Pa. (2)	30.18	30.18	30.19	30.85	30.90	30.90	29.35	29.35	29.35
Muscatine, Iowa	29.50	29.44	29.52	29.93	29.93	29.93	29.11	29.10	29.11
Nantucket, Mass.	30.12	30.11	30.10	30.86	30.83	30.82	29.28	29.31	29.15
Newark, N. J. (6)	30.01	......	30.02	30.79	......	30.78	29.23	......	29.65
New Bedford, Mass.	30.00	29.95	29.98	30.75	30.72	30.70	29.14	29.18	28.98
Newburybort, Mass.	30.04	30.02	30.07	31.07	30.88	30.83	29.23	29.16	29.09
New Harmony, Ia.	29.70	29.64	29.68	30.27	30.30	30.32	28.77	28.88	29.17
New Orleans, La.	30.29	30.26	30.31	30.65	30.61	30.86	29.85	29.89	29.86
New Wied, Texas	29.06	28.98	29.02	29.50	29.40	29.40	28.38	28.58	28.50
Norristown, Pa.	29.99	29.95	30.02	30.71	30.69	30.68	29.19	29.12	29.21

(1) Maximum for the month 30.71, min. 29.64.
(2) Not corrected for temperature.
(3) Obs. taken at 9 A. M. and 9 P. M.
(4) For the last 18 days of the month only.
(5) For the first 11 days of the month only.
(6) Obs. taken at 7 A. M. and 5 P. M.
(7) Obs. taken at Sunrise, Noon, and Sunset.

NAME OF STATION.	Mean.			Maxima.			Minima.		
	7 A. M.	2 P. M.	9 P. M.	7 A. M.	2 P. M.	9 P. M.	7 A. M.	2 P. M.	9 P. M.
Oberlin, Ohio	29.09	28.99	29.07	28.89	29.61	29.84	28.50	28.32	28.16
Oswego, N. Y. (1)	29.54	29.59	29.59	30.30	30.35	30.29	28.78	28.95	28.85
Oxford, Miss	29.62	29.57	29.60	30.06	30.07	30.12	29.04	29.12	29.06
Penn Yan, N. Y. (2)	29.19	29.23	29.24	30.00	29.98	29.95	28.39	28.59	28.52
Perrysburg, Ohio (3)	29.35	29.33	29.31	30.16	30.12	30.11	28.71	28.51	28.40
Philadelphia, Pa	29.98	29.95	29.98	30.58	30.61	30.57	29.27	29.16	29.20
Pittsburg, (Oak Sta.) Pa.	29.00	28.95	28.96	29.79	29.67	29.64	28.37	28.19	28.02
Pittsburg, (Marine Hospital,) Pa. (4)	28.98	28.96	28.24	29.78	29.67	29.57	28.40	28.25	28.05
Pomfret, Conn	28.96	28.93	28.28	29.70	29.66	29.64	28.14	28.27	28.03
Pottsville, Pa	29.31	29.25	29.30	30.05	30.03	29.91	28.12	28.32	28.63
Princeton, Mass	28.90	28.84	28.87	29.68	29.64	29.61	27.95	28.17	27.88
Providence, R. I. (5)	29.98	30.01	29.99	30.70	30.72	30.69	29.02	29.16	29.36
Quasqueton, Iowa	29.05	29.03	29.06	29.59	29.57	29.59	28.60	28.62	28.73
Sacramento, (Logan) California (6)	......	......	......	......	......	......	......	......	......
Sacramento, (Hatch) California (7)	......	......	......	......	......	......	......	......	......
Sag Harbor, N. Y.	29.83	29.82	29.86	30.57	30.58	30.59	29.02	29.04	28.91
St. Johnsburg, Vt.	29.39	29.37	29.34	30.03	30.10	30.05	28.30	28.46	28.50
St. Louis, Mo	29.49	29.46	29.46	30.16	30.16	30.19	28.64	28.90	28.99
St. Martin's, Can. E.	29.95	29.87	29.94	30.70	30.65	30.72	29.05	29.06	29.24
Savannah, Ga	30.15	30.09	30.13	30.56	30.52	30.61	29.58	29.47	29.55
Smithsonian Inst., D. C.	30.13	30.08	30.12	30.81	30.80	30.78	29.42	29.35	29.47
Southwick, Mass. (1)	29.99	29.94	29.88	30.83	30.72	30.44	29.12	29.24	29.06
Sparta, Ga	29.50	29.47	29.48	29.89	29.84	29.92	29.08	28.90	29.02
Spencertown, N. Y.	29.02	28.99	28.05	30.07	30.08	30.03	28.37	28.36	28.25
Springdale, Ky	29.42	29.40	29.42	30.10	30.15	30.11	28.69	28.52	28.66
Steuben, Me	30.16	30.12	30.15	30.87	30.82	30.87	29.23	29.21	29.31
The Rock, Ga. (8)	......	......	......	......	......	......	......	......	......
Thornburg, N. C.	30.13	30.07	30.10	30.75	30.73	30.74	29.36	29.40	29.39
Tuscaloosa, Ala	29.94	29.87	29.91	30.34	30.29	30.34	29.25	29.36	29.39
Upper Alton, Ill	29.45	29.42	29.46	30.09	30.07	30.14	28.67	28.89	28.99
Warrington, Fla.	30.10	30.10	30.06	30.44	30.43	30.40	29.57	29.60	29.61
Westfield, Mass.	30.08	30.08	30.08	31.47	31.60	31.45	30.87	30.87	30.84
Williamstown, Mass.	29.30	29.25	29.32	30.09	30.07	30.04	28.39	28.45	28.29
Woodward High School, Ohio (9)	29.44	29.45	29.45	30.17	30.06	30.19	28.81	28.37	28.79
Worcester, Mass	29.46	29.42	29.46	30.13	30.14	30.17	28.70	28.72	28.59
Zanesville, Ohio	29.20	29.17	29.16	29.85	29.70	29.77	28.66	28.29	28.39

(1) Incomplete.
(2) Obs. taken at Sunrise, Noon, and 10 P. M.; Mean for the month, 30.07; Max., 30.28; Min., 29.55.
(3) Obs. taken at 7 A. M., 1 P. M., and 9 P. M.
(4) Obs. at Sunrise, 9 A. M., 3 P. M., and 9 P. M.; Mean at 9 A. M., 29.01; Max., 29.80; Min., 28.41.
(5) Obs. taken at Sunrise, 1 P. M., and 10 P. M.
(6) Mean for the month, 29.95; Maximum, 30.34; Minimum, 29.44.
(7) Obs. taken at Sunrise, 2 P. M., and Sunset.
(8) Monthly Mean, 29.39; Max., 29.80; Min., 28.10; Obs. taken Morning, 3 P. M., and Evening.
(9) Obs. taken at 7 A. M., Noon, and 9 P. M.

NAME OF STATION.	Mean.			Maxima.			Minima.		
	7 A. M.	2 P. M.	9 P. M.	7 A. M.	2 P. M.	9 P. M.	7 A. M.	2 P. M.	9 P. M.
Alexandria, Va	30.00	29.96	30.02	30.33	30.26	30.28	29.57	29.59	29.67
All Saints, S. C	29.98	29.94	29.97	30.25	30.22	30.24	29.55	29.43	29.54
Amherst, Mass	29.67	29.64	29.67	30.15	30.12	30.12	29.33	29.29	29.38
Ann Arbor, Mich	29.04	29.01	29.03	29.45	29.42	29.42	28.75	28.65	28.68
Auburn, Ala	29.36	29.33	39.35	29.60	29.56	29.64	28.90	28.96	29.02
Ballardsville, Ky	29.08	29.05	29.02	29.41	29.37	29.41	28.66	28.55	28.62
Bedford, Pa	29.04	29.03	29.04	29.38	29.40	29.39	28.69	28.70	28.75
Beloit, Wis	29.23	29.23	29.24	29.56	29.64	29.68	28.66	28.76	28.73
Bloomfield, N. J(1)	29.77	29.74	29.77	30.18	30.14	30.21	29.40	29.40	29.45
Burlington, Vt	29.57	29.55	29.57	29.96	29.94	30.01	29.19	29.15	29.19
Burlington, N. J	29.98	29.98	29.99	30.22	30.30	30.31	29.70	29.64	29.70
Camden, S. C	29.87	28.93	29,85	30.19	30.12	30.17	29.41	29.46	29.43
Carmel, Me	29.54	29 44	29.50	30.04	29.99	29.91	29.17	29.08	29.12
Canton, N. Y	29.42	29.37	29.38	29.87	29.77	29.80	28.97	28.98	29.05
Cedar Keys, Fa	29.93	29,91	29.92	30.21	30.22	30.21	29.70	29.68	29.68
Ceresco, Wis	29.12	29.12	29.12	29.46	29.56	29.43	28.64	28.71	28.70
Chapel Hill, N. C	29.42	29,40	29.41	29.66	29.67	29.71	29.10	28.93	29.09
Charleston, S. C. (2)	....	....	....	....	....	....	....	....	....
Concord, N. H	29.55	29,49	29.54	30.05	30.04	30.05	29.19	29.15	29.21
Deaf &Dumb Inst.N. Y	29.89	29,89	29.90	30.25	30.24	30.26	29.52	29.53	29.56
Detroit, Mich	29.52	29,49	29.50	29.93	29.90	29.90	29.25	29.17	29.13
Dubuque, Iowa	29.38	29.35	29,39	29.77	29.79	29.85	28.83	28.77	28.77
Frederick City, Md	29.63	29.59	29.62	29.89	29.90	29.92	29.38	29.29	29.42
Gardiner, Me	29.73	29.69	29.74	30.23	30.20	30.21	29.41	29.36	29.38
Germantown, Ohio	29.27	29.21	29.24	29.64	29.60	29.71	28.84	28.71	28.75
Gettysburg, Pa	29.41	29.37	29.40	29.71	29.68	29.77	29.02	29.04	29.08
Glenwood, Tenn	29.62	29.57	29.61	29.92	29.90	29.88	29.07	29.05	29.10
Granville, Ohio	28.97	28.94	28.95	29.38	29.38	29.39	28.65	28.51	28.52
Great Falls, N.H	29.43	29.38	29.42	29.95	30.02	29.93	29.10	29.09	29.12
Hamilton, Canada. (3)	29.62	....	29.63	30.02	....	30.05	29.32	....	29.20
Harrisburg, Pa	29.72	29.69	29.71	30.00	30.02	30.09	29.33	29.37	29.40
Hillsborough, Ohio	28.87	28.83	28.81	29.23	29.22	29.28	28.50	28.35	28,36
Jacksonville, Fa	30.14	30.07	30.10	30.42	30.36	30.55	29.73	29.65	29.74
Key West, Fa	29.72	29.76	....	30.00	30.03	....	29.50	29.50	....
Knoxville, Tenn	29.03	28.98	29.03	29.48	29.31	29.38	28.57	28.56	28.59
Knox Hill, Fa.	29.89	29.80	29.88	30.12	30.10	30.16	29.51	29.55	29.55
Lima, Pa.	29.80	29.77	29.80	30.11	30.11	30.20	29.45	29.44	29.45
Madison, Ind	29.48	29.45	29.49	29.77	29.82	29.93	29.04	28.95	28.95
Manchester, N. H	29.93	29.89	29.92	30.45	30.48	30.51	29.64	29.65	29.73
Marietta, Ohio	29.25	29.22	29.23	29.60	29.58	29.66	28.84	28.79	28.81
Milton, Ind	29.07	29.04	29.09	29·42	29.44	29.51	28.64	28.48	28.55
Milwaukie, Wis	29.32	29.30	29.31	29.80	29.76	29.75	28.79	28.81	28.86
Monroe, Mich. (1)	28.83	28.83	28.80	29.30	29.40	29.30	28.20	28.50	28.40
Morrisville, Pa. (1)	30.06	30.04	30.06	30.40	30.40	30.45	29.65	29.70	29.70
Muscatine, Iowa	29.62	29.61	29.61	29.99	29.99	29.95	29.03	29.10	29.10
Nantucket, Mass	29.89	29.87	29.89	30.39	30.37	30.35	29.51	29.52	29.52
Newark, N. J, (4)	29.95	....	29.93	30.31	....	30.27	29.56	....	29.61
New Bedford, Mass	29.84	29.81	29.85	30.32	30.32	30.32	29.52	29.51	29.48
Newburyport, Mass	29.86	29.87	29.91	30.42	30.39	30.41	29.55	29.53	29.60
New Harmony, Ind	29.80	29.74	29.79	30.03	30.02	30.10	29 10	29.07	29.19
New Orleans, La	30.30	30.25	30.29	30.51	30.44	30.51	29.91	29.87	29.87
New Wied, Texas	29.09	28.98	29.03	29.45	29.40	29.40	28.60	28.60	28.60
Norristown, Pa	29.86	29.82	29.87	30.18	30.17	30.26	29.49	29.48	29.51
North Attleboro, Mass	29.71	29.67	29.71	30.23	30.14	30.21	29.38	29.34	29.37
Oberlin, Ohio	29.10	29.06	29.09	29.50	29.47	29.60	28.81	28.69	28.72
Orange Hill, Fa,	29.89	29.80	29.88	30.12	30.10	30.16	29.51	29.55	29.55

(1) Not corrected for temperature.
(2) Max. for the month 30.43 Min. 29.64
(3) Obs. taken at 9 o'clock A. M. and 9 P. M.
(4) Obs. taken at 7 A. M. and 5 P. M.

NAME OF STATION.	Mean.			Maxima.			Minima.		
	7 A.M.	2 P.M.	9 P.M.	7 A.M.	2 P.M.	9 P.M.	7 A.M.	2 P.M.	9 P.M.
Oswego, N. Y. (1)	29.47	29.50	29.50	29.84	29.88	29.90	29.14	29.24	29.25
Oxford, Miss.	29.62	29.58	29.59	29.89	29.91	29.96	29.21	29.08	29.13
Penn Yan, N. Y. (2)	29.12	29.15	29.16	29.50	29.50	29.50	28.79	28.86	28.85
Perrysburg, Ohio (3)	29.38	29.35	29.36	29.78	29.76	29.83	29.09	28.94	28.93
Philadelphia, Pa.	29.86	29.82	29.83	30.14	30.11	30.20	29.56	29.54	29.57
Pittsburg, Pa.:									
Oakland Station.	28.95	28.93	28.89	29.36	29.26	29.30	28.60	28.60	28.53
Marine Hospital (4)	28.94	28.93	28.94	29.30	29.30	29.40	28.50	28.50	28.60
Pomfret, Conn.	28.79	28.76	28.84	29.23	29.21	29.21	28.50	28.46	28.46
Pottsville, Pa.	29.20	29.15	29.17	29.47	29.47	29.47	28.84	28.84	28.88
Princeton, Mass.	28.67	28.66	28.69	29.20	29.18	29.17	28.31	28.30	28.39
Providence, R. I.	29.71	29.68	29.73	30.19	30.18	30.20	29.42	29.35	29.42
Quasqueton, Iowa	29.13	29.14	29.09	29.52	29.55	29.56	28.53	28.51	28.55
Sacramento, Cal.:									
Logan (5)	....	....	....	....	....	....	....	....	....
Hatch (6)	....	....	....	....	....	....	....	....	....
Sag Harbor, N. Y.	29.65	29.68	29.71	30.10	30.14	30.14	29.30	29.38	29.35
St. Louis, Mo.	29.54	29.52	29.54	29.87	29.90	30.00	28.83	28.87	28.89
St. Martin's, Canada E.	29.81	29.81	29.83	30.20	30.24	30.24	29.40	29.44	29.44
Savannah, Ga.	30.06	30.00	30.04	30.35	30.30	30.30	29.62	29.51	29.64
Smithsonian Inst., D. C.	30.06	29.97	30.02	30.35	30.29	30.39	29.64	29.58	29.67
Southwick, Mass. (7)	29.92	29.85	29.93	30.35	....	....	29.62	....	....
Sparta, Ga.	29.40	29.38	29.39	29.69	29.61	29.65	29.03	29.00	29.11
Spencertown, N. Y.	29.14	29.13	29.14	29.55	29.56	29.59	28.71	28.81	28.61
Springdale, Ky.	29.05	29.00	29.01	29.78	29.77	29.77	28.97	28.88	28.90
Steuben, Me.	29.94	29.90	29.93	30.49	30.45	30.49	29.53	29.49	29.49
The Rock, Ga. (8)	....	....	....	....	....	....	....	....	....
Thornbury, N. C.	30.01	29.94	29.95	30.32	30.25	30.35	29.54	29.51	29.55
Tuscaloosa, Ala.	29.89	29.86	29.87	30.17	30.16	30.26	29.44	29.45	29.40
Upper Alton, Ill.	29.52	29.48	29.53	29.84	29.87	29.96	28.86	28.65	29.03
Urbana, Ohio	28.76	28.74	28.74	29.16	29.13	29.16	28.42	28.31	28.30
Warrington, Fa.	30.08	30.09	30.04	30.29	30.34	30.29	29.67	29.71	29.71
Westfield, Mass	29.81	29.80	29.81	29.98	30.01	30.00	29.72	29.64	29.71
Williamstown, Mass	29.16	29.14	29.17	29.59	29.56	29.56	28.83	28.84	28.88
Woodward High School, Ohio (9)	29.50	29.49	29.51	29.85	29.88	29.83	29.10	29.04	29.14
Wood's Hole, Mass.	29.16	29.14	29.16	29.63	29.62	29.63	28.86	28.84	28.86
Worcester, Mass.	29.38	29.37	29.40	29.87	29.73	29.82	29.09	29.02	29.12
Zanesville, Ohio	29.19	29.16	29.16	29.55	29.51	29.61	28.87	28.79	28.78

(1) Not corrected for temperature.
(2) Obs. taken at sunrise, 2 P. M., and sunset.
(3) Obs. taken at 7 A. M., 1 P. M., and 9 P. M.
(4) Obs. taken at sunrise, 9 A. M., 3 P. M., and 9 P. M. Mean at 9 A. M., 28.96; max., 29.30; min., 28.50.
(5) Mean for the month, 29.78; max., 30.11; min., 28.50.
(6) Obs. taken at sunrise, noon, and 10 P. M. Mean for the month, 30.05; max., 30.29; min., 29.69.
(7) Obs. incomplete.
(8) Mean for the month, 29.37; max., 29.60; min., 29.25. Obs. taken morning, 3 P. M., and evening.
(9) Obs. taken at 7 A. M., noon, and 9 P. M.

NAME OF STATION.	Mean.			Maxima.			Minima.		
	7 A. M.	2 P. M.	9 P. M.	7 A. M.	2 P. M.	9 P. M.	7 A. M.	2 P. M.	9 P. M.
Alexandria, Va	29.99	29.91	29.94	30.47	30.40	30.42	29.44	29.36	29.53
All Saints, S. C.	29.99	29.94	29.96	30.34	30.30	30.33	29.65	29.36	29.23
Amherst, Mass.	29.64	29.60	29.61	30.14	30.08	30.09	29.06	29.04	29.08
Ann Arbor, Mich. (Winchell,)	28.96	28.93	28.97	29.49	29.45	29.35	28.44	28.31	28.52
Auburn, Ala.	29.37	29.33	29.34	29.70	29.68	29.68	29.07	29.07	29.08
Baldwin Inst, Ohio	28.88	28.87	28.91	29.52	29.55	29.45	28.34	28.31	28.39
Ballardsville, Ky	29.05	29.04	29.01	29.50	29.50	29.45	28.37	28.51	28.62
Bedford, Pa.	29.02	28.98	28.99	29.50	29.50	29.48	28.47	28.63	28.55
Beloit, Wis.	29.11	29·10	29.11	29.62	29.56	29.56	28.43	28.37	28.63
Bloomfield, N. J.	29.76	29.70	29.75	30.22	30.22	30.19	29.22	29.27	29.27
Burlington, N. J. (1)	29.97	29.92	29.93	30.42	30.39	30.47	29.47	29.46	29.56
Burlington, Vt.	29.50	29.48	29.49	30.07	29.96	29.94	28.77	28.86	28.87
Camden, S. C.	29.89	29.84	29.85	30.29	30.23	30.26	29.47	29.50	29.42
Canton, N. Y. (2)	29.29	29.27	29.23	29.85	29.90	29.91	28.73	28.74	28.63
Carmel, Me.	29.43	29.40	29.46	29.99	29.90	29.92	28.83	28.81	28.98
Cedar Keys, Fa.	29.91	29.89	29.91	30.23	30.18	30.19	28.99	28.93	29.66
Ceresco, Wis.	28.96	28.98	28.98	29.47	29.45	29.34	28.17	28.19	28.20
Chapel Hill, N. C.	29.43	29.38	29.39	29.81	29.72	29.81	29.06	28.95	28.91
Charleston, S. C. (3)	....	....	....	....	....	....	....	....	....
Concord, N. H.	29.49	29.46	29.46	29.97	29.93	29.90	28.87	28.95	28.98
Deaf & Dumb Inst. N. Y.	29.94	29.85	29.87	30.36	30.32	30.33	29.35	29.35	29.25
Detroit, Mich.	29.42	29.41	29.43	29.99	29.93	29.97	28.93	28.76	28.96
Dubuque, Iowa	29.27	29.25	29.28	29.76	29.71	29.43	28.52	28.64	28.76
Frederick, Md.	29.55	29.48	29.50	30.01	29.94	29.95	29.01	29.08	29.11
Gardiner, Me.	29.68	29.65	29.68	30.11	30.09	30.08	29.10	29.09	29.22
Germantown, O.	29.20	29.17	29.16	29.77	29.69	29.61	28.49	28.57	28.75
Gettysburg, Pa.	29.40	29.34	29.35	29.87	29.84	29.83	28.79	28.89	28.90
Glenwood, Ind.	29.60	29.53	29.54	30.07	29.99	30.01	29.02	28.96	28.87
Granville, Ohio	28.94	28.91	28.90	29.44	29.46	29.40	28.38	28.23	28.50
Great Falls, N. H.	29.37	29.35	29.37	29.89	29.86	29.80	28.77	28.78	28.87
Hamilton, C. E.	29.57	....	29.55	30.04	....	30.00	29.06	....	29.00
Harrisburg, Pa.	29.70	29.64	29.67	30.17	30.16	30.12	29.11	29.18	29.22
Hillsborough, Ohio	28.83	28.80	28.81	29.33	29.28	29.23	28·21	28.15	28.42
Jacksonville, Fa.	30.14	30.07	30.10	30.43	30.41	30.38	29.56	29.57	29.45
Key West, Fa.	29.71	29.74	....	29.88	29.92	....	29.42	29.43	....
Knox Hill, Fa.	29.89	29.85	29.86	30.24	30.17	30.22	29.60	29.56	29.61
Knoxville, Tenn.	29.03	28.98	29.01	29.48	29.40	29.44	28.58	28.56	28.66
Lebanon, Tenn.	29.54	29.46	29.46	29.95	29.92	29.85	29.19	29.02	28.98
Lima, Pa.	29.80	29.73	29.76	30.28	30.23	30.23	29.23	29.20	29.32
Madison, Ind.	29.48	29.46	29.43	29.97	29.94	29.86	28.77	28.84	28.99
Manchester, N. H.	29.91	29.87	29.89	30.37	30.30	30.29	29.40	29.28	29.44
Marietta, Ohio	29.21	29.17	29.19	29.68	29.61	29.60	28.70	28.73	28.80
Milton, Ind.	29.02	28.98	29.05	29.45	29.45	29.47	28.24	28.34	28.16
Milwaukie, Wis.	29.21	29.19	29.23	29.71	29.62	29.57	28.58	28.40	28.68
Morrisville, Pa. (1)	30.07	30.01	30.01	30.50	30.50	30.50	29.50	29.45	29.60
Muscatine, Iowa	29.45	29.44	29.47	29.90	29.90	29.90	28.75	28.90	29.06
Nantucket, Mass.	29.91	29.85	29.86	30.35	30.32	30.34	29.42	29.32	29.45
Newark, N. J. (4)	29.95	....	29.81	30.40	....	30 35	29.38	....	29.50
New Bedford, Mass.	29.84	29.79	29.82	30.27	30.22	30.24	29.27	29.13	29.36
Newburyport, Mass.	29.85	29.81	29.82	30.38	30.31	30.29	29.24	29.20	29.38
New Harmony, Ind.	29.70	29.64	29.65	30.17	30.08	30.10	28.94	29.07	29.05
New Orleans, La	30.27	30.21	30.24	30.71	30.54	30.63	29.93	29.88	29.88
New Wied, Texas. (1)	29.01	28.89	28.86	29.52	29.30	29.30	28.46	28.48	28.48
Norristown, Pa.	29.86	29.78	29.80	30.35	30.27	30.27	29.20	29.16	29.36
North Attleboro, Mass.	29.70	29.64	29.66	30.18	30.12	30.15	29.14	29.06	29.23

* For the changes in hours of observations see tables of Thermometer for this month.
(1) Not corrected for temperature.
(2) Obs. omitted from the 7th. to the 16th. of the month.
(3) Maximum for the month 30.47, Min. 29.44.
(4) Obs. taken at 7 A. M. and 5 P. M.

NAME OF STATION.	Mean.			Maxima,			Minima.		
	7 A. M.	2 P. M.	9 P. M.	7 A. M.	2 P. M.	9 P. M.	7 A. M.	2 P. M.	9 P. M.
Oberlin, Ohio.	29.03	28.96	29.03	29.58	29.37	29.48	28.48	28.37	28.60
Oswego, N. Y. (1)	29.47	29.44	29.41	29.90	29.90	29.85	28.94	28.95	28.88
Oxford, Miss.	29.59	29.55	29.53	30.02	29.96	29.95	29.15	29.08	29.07
Penn Yan, N. Y.	29.10	29.10	29.10	29.57	29.60	29.55	28.55	28.56	28.55
Perrysburg, Ohio. (2)...	29.32	29.26	29.30	29 87	29.79	29.71	28.69	28.63	28.70
Philadelphia, Pa.	29.83	29.77	29.79	30.27	30.23	30.25	29.37	29.31	29.41
Pittsburg, Pa. (Oakland Station.)	28.93	28.90	28.91	29.45	29.49	29.35	28.46	28.39	28.58
Pittsburg, Pa. (Marine Hospital.) (3)..........	28.90	28.87	28.90	29.40	29.40	29.40	28.44	28.45	28.50
Pomfret, Conn.	28.79	28.75	28.76	29.22	29.20	29.18	28.27	28.21	28.36
Pottsville, Pa.	29.19	29.12	29.16	29.67	29.58	29.58	28.56	28.65	28.67
Princeton, Mass.	28.67	28.66	28.65	29.10	29.10	29.06	28.11	28.08	28.17
Providence, R. I.	29.69	29.63	29.66	30.18	30.12	30.14	29.26	29.08	29.21
Quasqueton, Iowa.	29.02	29.02	29.03	29.49	29.49	29.47	28.31	28.35	28.58
Sacramento, (Logan) ... California. (4)...........	....	....	....	....	....	....	....	....	....
Sacramento, (Hatch) ... California. (5)...........	....	....	....	....	....	....	....	....	....
Sag Harbor, N. Y.	29.63	29.61	29.64	30.10	30.06	30.01	29.12	29.05	29.17
St. Johnsbury, Vt.	29.10	29.06	29.08	29.59	29.47	29.54	28.34	28.70	28.52
St. Louis, Mo.	29.51	29.46	29.47	30.03	29.97	29.96	28.79	28.89	28.78
St. Martins, C. E.	29.74	29.71	29.70	30.19	30.10	30.10	29.10	29.21	29.04
San Francisco, Cal. (6)..	29.83	29.79	29.83	....	....	....	....	....	....
Savannah, Ga.	30.08	30.00	30.04	30.42	30.36	30.42	29.71	29 40	29.59
Smithsonian Inst.	30.00	29.90	29.95	30.51	30.28	30.42	29.38	29.36	29.52
Southwick, Mass. (7).....	29.91	29.78	29.82	30.51	30.06	29.94	29.29	29.31	29.71
Sparta, Ga.	29.44	29.39	29.42	29.75	29.73	29.79	29.17	29.08	29.09
Spencertown, N. Y.	29.09	29.08	29.07	29.66	29.63	29.59	28.56	28.67	28.64
Springdale, Ky.	29.40	29.37	29.37	29.90	29.84	29.78	28.67	28.84	28.88
Steuben, Me.	29.86	29.82	29.84	30.36	30.20	30.25	29.25	29.22	29.27
The Rock, Ga. (8).........	....	....	....	....	....	....	....	....	....
Thornbury, N. C.	30.01	29.94	29.96	30.44	30.37	30.42	29.59	29.56	29.44
Tuscaloosa, Ala.	29.90	29.83	29.83	30.35	30.25	30.24	29.50	29.43	29.32
Upper Alton, Ill.	29.51	29.45	29.45	29.96	29.84	29.82	28.80	28.95	28.33
Urbana, Ohio.	28.70	28.70	28.71	29.12	29.23	29.14	28.11	28.07	28.31
Waccaman, S. C.	29.99	29.94	29.96	30.34	30.30	30.33	29.65	29.36	29.23
Warrington, Fa. (9)......	30.06	30.07	30.00	30.48	30.36	30.38	29.75	29.79	29.73
Westfield, Mass.	29.80	29.80	29.80	29.96	29.98	29.98	29.71	29.66	29.69
Williamstown, Mass.	29.15	29.14	29.16	29.64	29.59	29.56	28.61	28.60	28.62
Woods Hole, Mass.	29.13	29.12	29.15	29.59	29.59	29.56	28.64	28.55	28.67
Worcester, Mass.	29.37	29.33	29.35	29.80	29.73	29.75	28.85	28.93	28.93
Zanesville, Ohio.	29.14	29.08	29.12	29.65	29.55	29.56	28.65	28.52	28.70

(1) Not corrected for temperature. (2) Obs. taken at 7 A. M. 1 P. M. and 9 P. M.
(3) Obs. taken at sunrise, 9 A. M. 3 P. M. and 9 P. M. Mean at 9 A. M. 28.91 Max. 29.42. Min. 28.44. (4) Mean for the month 29.98. Max. 30.45. Min. 29.00.
(5) Obs. taken at sunrise, noon and 10 P. M. Mean for the month 30.01. Max. 30.16. Min. 29.72. (6) Obs. taken at sunrise, 9 A. M. noon and 10 P. M. Mean at 9 A. M. Maximum29.835. for the month 29.96. Minimum. 29.50.
(7) Obs. irregular, none at 9 A. M. except on the 13th. and 15th.
(8) Mean for the month 29.46. Max. 29.65. Min. 29.35. Obs. taken morning 3 P. M. & ev ening.
(9) Obs. taken at sunrise, noon and sunset.

NAME OF STATION.	Mean.			Maxima.			Minima.		
	7 A. M.	2 P. M.	9 P. M.	7 A. M.	2 P. M.	9 P. M.	7 A. M.	2 P. M.	9 P. M.
Alexandria, Va	30.04	29.99	30.04	30.36	30.28	30.29	29.46	29.43	29.67
All Saints, S. C	30.05	30.04	30.03	30.27	30.24	30.24	29.66	29.61	29.65
Amherst, Mass	29.75	29.73	29.73	30.03	30.00	30.06	29.09	28.80	29.02
Ann Arbor, Mich	28.08	29.06	29.06	29.46	29.39	29·43	28.81	28.81	28.79
Auburn, Ala	29.41	29.36	29.37	29.60	29.54	29.52	29.17	29.09	29.13
Bedford, Pa	29.15	29.10	29.12	29.43	29.42	29.37	28.82	28.85	28.82
Beloit, Wis	29.21	29.17	29.17	29.66	29.57	29.63	28.90	28.92	28.87
Bloomfield, N. J	29.86	29.83	29.87	30.19	30.11	30.12	29.13	29.05	29.33
Burlington, N. J	29.99	29.94	29.97	30.32	30.21	30.22	29.46	29.34	29.53
Burlington, Vt	29.61	29.57	29.59	29.94	29.94	29.93	28.86	28.80	28.84
Camden, S. C	29.96	29.91	29.92	30.19	30.11	30.15	29.57	29.54	29.56
Canton, N. Y	29.42	29.37	29.41	29.78	29.73	29.81	28.85	28.85	28.86
Carmel, Me	29.57	29.49	29.56	29.93	29.88	29.99	28.67	28.57	28.39
Cedar Keys, Fa	29.96	29.96	29.95	30.13	30.13	30.13	29.65	29.70	29.74
Ceresco, Wis	29.09	29.07	29.07	29.54	29.59	29.55	28.73	28.75	28.71
Chapel Hill, N. C	29.51	29.47	29.49	29.73	29.70	29.71	28.97	29.02	29.11
Charleston, S. C. (1)	....	....	....	....	....	....	....	....	....
Concord, N. H	29.61	29.57	29.59	29.93	29.92	29.88	28.84	28.67	28.72
Deaf & Dumb Inst. N. Y.	29.97	29.93	29.96	30.27	30.21	30.19	29.29	29.20	29.48
Detroit, Mich	29.53	29.52	29.52	29.91	29.87	29.89	29.24	29.25	29.22
Dubuqne, Iowa	29.35	29.28	29.31	29.78	29.71	29.77	28.97	29.00	28.98
Frederick, Md	29.60	29.54	29.58	29.88	29.87	29.80	29.12	29.13	29.26
Gardiner, Me	29.79	29.75	29.77	30.10	30.09	30.10	28.95	28.84	28.73
Germantown, Ohio	29.28	29.22	29.26	29.94	29.51	29.56	29.00	29.00	29.03
Gettysburg, Pa	29.47	29.40	29.44	29.76	29.66	29.67	28.89	28·91	29.08
Glenwood, Tenn	29.61	29.54	29.58	29.89	29.84	29.94	29.39	29.31	29.36
Granville, Ohio	29.00	28.94	28.96	29.41	29.30	29.30	28.65	28.70	28.71
Great Falls, N. H	29.51	29.47	29.50	29.83	29.83	29.79	28.61	28.44	28.56
Hamilton, Ca	29.68	....	29.70	30.00	....	30.03	29.32	....	29.38
Harrisburg, Pa	29.77	29.71	29.75	30.07	29.98	29.96	29.20	29.20	29.38
Hillsborough, Pa	28.92	28.89	28.89	29.28	29.25	29.16	28.63	28.61	28.65
Jackson, Ohio. (Gil-	....	....	....	....	....	....	....	....	....
mor) (2)	28.97	29.05	28.99	29.33	29.83	29.50	28.33	28.42	28.25
Jackson, O. (Wood) (2)	29.00	29.06	29.03	29.37	29.37	29.25	28.29	28.83	28.83
Jacksonville, Fa	30.17	30.14	30.16	30.33	30.36	30.36	29.86	29.87	29.85
Key West, Fa	29.76	29.81	....	29.91	29.95	....	29.58	29.58	....
Knox Hill, Fa	29.58	29.55	29.55	29.79	29.77	29.77	29.39	29.33	29.35
Knoxville, Tenn	29.11	29.05	29.07	29.36	29.28	29.32	28.81	28.74	28.84
Lima, Pa	29.88	29.81	29.86	30.17	30.12	30.10	29.21	29.18	29.44
Madison, Ind	29.53	29.52	29.51	29.89	29.86	29.84	29.35	29.28	29.31
Manchester, N. H	29.96	29.92	29.95	30.30	30.26	30.26	29.32	29.05	29.13
Marietta, Pa	29.31	29.27	29.29	29.69	29.57	29.57	28.92	28.98	28.93
Milton, Ind	29.10	29.09	29.09	29.40	29.43	29.40	28.82	28.86	28.86
Milwaukie, Wis	29.31	29.29	29.32	29.74	29.65	29.73	28.99	28.98	29.98
Morrisville, Pa. (2)	30.13	30.10	30.14	30.45	30.45	30.40	29.50	29.50	29.70
Muscatine, Iowa	29.55	29.50	29.51	29.92	29.88	29.87	29.23	29.23	29.23
Nantucket, Mass	29.94	29.94	29.96	30.31	30.31	30.27	29.06	28.89	29.22
Newark, N. J	30.00	....	29.99	30.30	....	30.36	29.30	....	29.36
New Bedford, Mass	29.88	29.84	29.87	30.21	30.19	30.17	29·18	28.87	29.13
Newburyport, Mass	29.96	29.88	29.94	30.29	30.23	30.26	29.35	28.94	29.02
New Orleans, La	30.26	30.22	30.25	30.40	30.37	30.40	30.03	30.01	30.02
New Wied, Tex. (2)	28.85	28.75	28.77	29.10	29.00	29.05	28.62	28.55	28.52
Norristown, Pa	29.96	29.92	29.95	30.28	30.22	30.43	29.46	29.52	29.65
North Attleboro, Mass.	29.80	29.74	29.80	30.13	30.07	30.09	29.15	28.79	29.04
Oberlin, Ohio	29.13	29.11	29.20	29.51	29 47	29.44	28.88	28.87	28.85
Oxford, Miss	29.58	29.55	29.55	29.79	29.77	29.77	29.39	29.33	29.35

* For changes in hours of observations, see tables of Thermometer for this month.

(1) Maximum for the month, 30.44, Min. 29.85.

(2) Not corrected for temperature.

NAME OF STATION.	Mean.			Maxima.			Minima.		
	7 A. M.	2 P. M.	9 P. M.	7 A. M.	2 P. M.	9 P. M.	7 A. M.	2 P. M.	9 P. M.
Penn Yan, N. Y.	29.21	29.24	29 27	29.55	29.56	29.64	28.70	28.80	28.80
Perrysburg, Ohio	29.43	29.37	29.40	29.78	29.73	29.73	29.15	29.06	29.07
Philadelphia, Pa.	29.88	29.82	29.86	30.18	30.10	30.08	29.34	29.23	29.49
Pittsburg, Penn. (Oakland Station.)	29.00	28.97	29.00	29.39	29.32	29.31	28.59	28.69	28.65
Pittsburg, Penn. (Marine Hospital.) (1)	28.96	28.96	28.98	29.35	29.31	29.33	28.56	28.69	28.65
Pomfret, Conn.	28.90	28.85	28.89	29.18	29.17	29.13	28.19	28.00	28.18
Pottsville, Pa.	29.25	29.19	29.25	29.57	29.47	29.50	28.63	28.62	28.81
Princeton, Mass.	28.80	28.75	28.80	29.10	29.08	29.58	28.05	27.84	27.96
Providence, R. I.	29.78	29.71	29.78	30.14	30.09	30.11	29.09	28.73	29.05
Quasqueton, Iowa.	29.09	29.04	29.04	29.53	29.46	29.50	28.72	28.71	28.73
Sag Harbor, N. Y.	29.74	29.67	29.76	30.07	30.07	30.07	29.05	28.80	29.07
St. Johnsbury, Vt.	29.20	29.16	29.19	29.52	29.50	29.53	28.50	28.46	28.28
St. Louis, Mo.	29.52	29.46	29.46	29.92	29.80	29.85	29.28	29.24	29.24
St. Martins, Ca. E.	29.83	29.85	29.86	30.21	30.19	30.16	29.04	29.19	29.19
Savannah, Ga.	30.12	30.08	30.10	30.34	30.30	30.31	29.80	29.76	29.76
Smithsonian Inst.	30.06	29.99	30.01	30.38	30.28	30.27	29.48	29.45	29.69
Southwick, Mass. (2)	29.95	29.92	30.00	30.27	30.22	30.24	29.35	29.12	29.84
Sparta, Ga.	29.46	29.42	29.45	29.65	29.60	29.62	29.20	29.17	29.21
Spencertown, N. Y.	29.26	29.21	29.26	29.56	29.52	29 54	28.56	28·50	28.62
Steuben, Me.	30.00	29.92	29.93	30.35	30.41	30.31	29.08	28.92	28.66
The Rock, Ga. (3)	....	....	....	....	....	....	....	....	....
Thornbury, N. C.	30.09	30.01	30.04	30.35	30.27	30.29	29.52	29.48	29.60
Upper Alton, Ill.	29.51	29.45	29.45	29.87	29.77	29.81	29.28	29.23	29.24
Warrington, Fa.	30.05	30.08	30.02	30.22	30.26	30.23	29.83	29.82	29.74
Westfield, Mass.	29.49	29.46	29.49	29.72	29.69	29.73	29.25	29.26	29.22
Williamstown, Mass.	29.25	29.18	29.25	29.51	29.47	29.48	28.62	28.66	28.80
Wood's Hole, Mass.	29.21	29.20	29.21	29.49	29.53	29.51	28.52	28.20	28.39
Worcester, Mass.	29.46	29.41	29.46	29.77	29.72	29.79	28.80	28.57	28.75
Zanesville, Ohio.	29.17	29.13	29.16	29.54	29.44	29.50	28.84	28.92	28.90

(1) Obs. taken at Sunrise, 9 A. M. 3 P.M. and 9 P. M. Mean at 9 A. M. 28.98, Max. 29.34, Min. 28.60

(2) Obs. irregular, None at 9 P. M. except on the 5th, 26th, and 28th.

(3) Mean for the month 29.80, Maximum 30.15, Min. 29.50, Obs. taken at morn. 3 P. M. & eve.

NAME OF STATION.	Mean.			Maxima.			Minima.		
	7 A. M.	2 P. M.	9 P. M.	7 A. M.	2 P. M.	9 P. M.	7 A. M.	2 P. M.	9 P. M.
Alexandria, Va	29.99	29.95	29.96	30.25	30.20	30.22	29.61	29.64	29.62
All Saints, S. C	29.95	29.94	29.94	30.18	30.15	30.16	29.60	29.63	29.70
Amherst, Mass	29.72	29.66	29.69	30.04	30.00	29.98	29.42	29.30	29.35
Ann Arbor, Mich	29.10	29.06	29.07	29.36	29.31	29.29	28.86	28.82	28.81
Auburn, Ala	29.32	29.27	29.28	29.51	29.45	29.44	29.12	29.12	29.06
Bedford, Pa	29.10	29·07	29.06	29.30	29.28	29.24	28.82	28·74	29.70
Beloit, Wis	29.22	29.19	29.19	29.53	29.43	29.47	28.81	28.78	28.79
Bloomfield, N. J	29.81	29.77	29.81	30.13	30.08	30.11	29.44	29.43	29.48
Burlington, N. J	29.91	29.86	29.90	30.16	30.16	30.13	29.60	29.59	29.60
Burlington, Vt	29.64	29.61	29.62	29.90	29.92	29.88	29.33	29.23	29.25
Camden, S. C	29.87	29.83	29.84	30.10	30.02	30.05	29.55	29.53	29.58
Canton, N. Y	29.46	29.42	29.42	29.70	29.69	29.68	29.13	29.13	29.13
Carmel, Me. (1)	29.56	29.53	29.55	29.83	29.77	29.81	29.27	29.21	29.24
Cedar Keys, Fa	29.86	29.85	29.85	30.01	29.99	30.00	29.69	29.67	29.67
Ceresco, Wis	29.17	29.12	29.12	29.46	29.37	29.45	28.74	28.69	28.57
Chapel Hill, N. C	29.43	29.41	29.42	29.61	29.60	29.64	29.11	29.14	29.14
Charleston, S. C. (2)	....	....	....	....	....	....	....	....	....
Concord, N. H	29.60	29.56	29.59	29.90	29.84	29.82	29.28	29.25	29.22
Deaf & Dumb Inst. N. Y	29.93	29.90	29.91	30.22	30.20	30.17	29.55	29.53	29.57
Detroit, Mich	29.55	29.52	29.52	29.78	29.78	29.74	29.31	29.28	29.22
Dubuque, Iowa	29.35	29.26	29.33	29.67	29.54	29.57	28.95	28.80	28.85
Frederick, Md	29.56	29.50	29.52	29.77	29.71	29.73	29.24	29.22	29.24
Gardiner, Me	29.76	29.74	29.76	30.11	30.01	30.03	29.46	29.44	29.45
Germantown, Ohio	29.23	29.18	29.20	29.48	29.44	29.41	28.96	28.97	28.95
Gettysburg, Pa	29.41	29.38	29.38	29.65	29.63	29.58	29.04	29.08	29.08
Glenwood, Tenn	29.53	29.48	29.50	29.72	29.68	29.70	29.31	29.21	29.24
Granville, Ohio	28.98	28.95	28.97	29.17	29.16	29.15	28.71	28.61	28.76
Great Falls, N. H	29.49	29.46	29.49	29.77	29.74	29.74	29.14	29.19	29.18
Hamilton, Canada	29.74	....	29.71	29.98	....	29.90	29.40	....	29.40
Harrisburg, Pa	29.73	29.69	29.70	29.98	29.93	29.92	29.34	29.40	29.42
Hillsborough, Ohio	28.89	28.85	28.85	29.10	29.09	29.06	28.64	28.47	28.60
Jackson, Ohio, (3)	28.98	28.98	28.95	29.17	29.25	29.17	28.75	28.79	28.62
Jacksonville, Fa	30.07	30.02	30.04	30.25	30.22	30.27	29.84	29.64	29.83
Key West, Fa	29.67	29.72	....	29.78	29.92	....	29.60	29.64	....
Knox Hill, Fa	29.81	29.78	29.79	29.97	29.93	29.93	29.61	29.61	29.60
Knoxville, Tenn	29.04	28.97	29.00	29.20	29.15	29.15	28.85	28.70	28.74
Lima. Pa	29.82	29.77	29.78	30·13	30.07	30.09	29.38	29.47	29.46
Madison, Wis	29.02	29.02	29.01	29.33	29.29	29.26	28.67	28.57	28.57
Madison, Ind	29.47	29.43	29.44	29.71	29.70	29.67	29.20	29.07	29.18
Madrid, N. Y	29.72	29.62	29.67	29.96	29.83	29.87	29.36	29.34	29.38
Manchester, N. H	29.97	29.92	29.95	30.29	30.17	30.18	29.64	29.60	29.62
Marietta, Ohio	29.30	29.27	29.28	29.50	29.49	29.47	29.08	28.94	28.95
Milton, Ind	29.07	29.01	29.05	29.31	29.27	29.30	28.97	28.68	28.64
Milwaukie, Wis	29.34	29.32	29.30	29.61	29.60	29.54	28.95	28.98	29.00
Morrisville, Pa. (3)	30.08	30.06	30.08	30.40	30.35	30.35	29.70	29.75	29.57
Muscatine, Iowa	29.50	29.48	29.49	29.77	29.74	29.74	29.19	28.99	29.17
Nantucket, Mass	29.92	29.89	29.92	30.30	30.25	30.28	29.39	29.35	29.30
Newark, N. J	29.93	....	29.93	30.17	....	30.10	29.64	....	29.60
New Bedford, Mass	29.86	29.82	29.86	30.21	30.22	30.15	29.45	29.37	29.46
Newburyport, Mass	29.95	29.89	29.92	30.23	30.19	30.18	29.59	29.56	39.57
New Orleans, La	30.16	30.13	30.14	30.27	30.24	30.27	30.04	29.99	30.01
New Wied, Texas. (3)	28.69	28.64	28.66	28.90	28.76	28.80	28.51	28.50	28.50
Norristown, Pa	29.87	29.84	29.87	30.18	30.11	30.13	29.45	29.55	29.59
North Attleboro, Mass	29.75	29.72	29.77	30.11	30.03	30.05	29.35	29.35	29.40
Oberlin, Ohio	29.12	29·11	29.10	29.36	29.39	29.32	28.89	28.87	28.91
Oswego, N. Y	29.63	29.64	29.63	29.82	29.85	29.77	29.38	29.37	29.36

* For changes in hours of observations, see tables of Thermometer for this month.

(1) No obs. for the first 14 days.

(2) Maximum for the month 30.39, Min. 29.84.

(3) Not corrected for temperature.

NAME OF STATION.	Mean.			Maxima.			Minima.		
	7 A. M.	2 P. M.	9 P. M.	7 A. M.	2 P. M.	9 P. M.	7 A. M.	2 P. M.	9 P. M.
Oxford, Miss	29.52	29.48	29.48	29.69	29.65	29.65	29.35	29.27	29.26
Penn Yan, N. Y	29.25	29.27	29.26	29.42	29.45	29.40	28.90	28.99	29.03
Perrysburg, Ohio	29.39	29.38	29.38	29.65	29.64	29.61	29.11	29.08	29.09
Philadelphia, Pa	29.83	29.78	29.81	30.12	30.06	30.10	29.45	29.49	29.51
Pittsburg, Pa. (Oakland Station.)	28.95	28.91	28.91	29.16	29.14	29.09	28.70	28.67	28.61
Pittsburg, Pa. (Marine Hospital.) (1)	28.95	28.94	28.94	29.17	29.11	29.13	28.63	28.69	28.64
Pomfret, Conn	28.88	28.85	28.86	29.20	29.16	29.15	28.57	28.51	28.56
Pottsville, Pa	29.22	29.19	29.20	29.49	29.45	29.43	28.86	28.91	28.92
Princeton, Mass	28.79	28.78	28.80	29.11	29.07	29.07	28.48	28.45	28.46
Providence, R. I	29.74	29.72	29.75	30.05	30.05	30.08	29.37	29.38	29.40
Quasqueton, Iowa	29.09	29.05	26.06	29.39	29.33	29.33	28.63	28.66	28.60
Sag Harbor, N. Y	29.68	29.64	29.61	30.02	29.96	30.00	29.24	29.24	29.30
St. Johnsbury, Vt	29.20	29.15	29.16	29.45	29.39	29.36	28.90	28.85	28.90
St. Louis, Mo	29.43	29.40	29.41	29.70	29.66	29.65	29.11	29.03	29.05
St. Martins, C. E	29.93	29.93	29.93	30.20	30.19	30.20	29.58	29 64	29.67
Savannah, Ga	30.02	29.96	30.01	30.26	30.21	30.23	29.73	29.07	29.76
Smithsonian Inst	30.02	29.96	29.98	30.29	30.23	30.23	29.64	29.66	29.61
Southwick, Mass(2)	29.95	29.92	29.90	30.22	30 21	30.03	29.72	29.62	29.69
Sparta, Ga	29.39	29.35	29.37	29.56	29.60	29.58	29.14	29.19	29.12
Spencertown, N. Y	29.25	29.23	29.24	29.53	29.49	29.47	29.01	28.90	28.86
Springdale, Ky	29.38	29.36	29.35	29.59	29.64	29.54	29.09	29.02	29.06
Steuben, Me	29.99	29.96	29.99	30.27	30.24	30.31	29.57	29.54	29.57
The Rock, Ga. (3)	....	....	....	....	....	....	....	....	....
Upper Alton, Ill	29.44	29.41	29.42	29.83	29.65	29.69	29.04	29.06	29.06
Urbana Ohio	28.77	28.75	28.76	28.99	28.99	29.01	28.55	28.46	28.52
Warrington, Fa	29.94	29.96	29.90	30.08	30.15	30.07	29.74	29.78	29.71
Westfield, Mass	29.77	29.76	29.78	29.97	29.94	29.94	29.67	29.66	29.69
Williamstown, Mass	29.24	29.21	29.44	29.58	29.46	29.24	28.95	28.97	29.01
Worcester, Mass	29.45	29.42	29.45	29.75	29.78	29.67	29.13	29.09	29.21

(1) Obs. taken at Sunrise, 9 A. M. 3 P. M. and 9 P. M. Mean at 9 A.M. 28.97, Max. 29.16, Min. 28.72.

(2) Obs. irregular, only four at 9 P. M.

(3) Mean for the month 29.83, Max. 30.00, Min. 29.45, obs. taken at Morn., 3 P. M. and Eve.

NAME OF STATION.	Mean.			Maxima.			Minima.		
	7 A. M.	2 P. M.	9 P. M.	7 A. M.	2 P. M.	9 P. M.	7 A. M.	2 P. M.	9 P. M.
Alexandria, Va.	29.93	29.90	29.91	30.20	30.13	30.14	29.64	29.57	29.62
All Saints, S. C.	29.97	29.96	29.96	30.20	30.20	30.20	29.75	29.70	29.65
Amherst, Mass.	29.66	29.61	29.64	29.95	29.92	29.92	29.23	29.28	29.13
Ann Arbor, Mich.	28.97	28.95	28.96	29.31	29.26	29.24	28.46	28.68	28.61
Auburn, Ala.	29.32	29.28	29.29	29.50	29.41	29.45	29.13	29.07	29.11
Bedford, Pa.	29.05	29.02	29.02	29.31	29.28	29.23	28.80	28.65	28.80
Beloit, Wis.	29.10	29.08	29.10	29.44	29.41	29.41	28.87	28.65	28.69
Bloomfield, N. J.	29.75	29.71	29.74	29.99	30.01	30.02	29.36	29.40	29.30
Burlington, Vt.	29.50	29.48	29.50	29.88	29.87	29.88	29.07	29.05	29.13
Burlington, N. J.(1)	29.90	29.89	29.88	30.15	30.15	30.10	29.60	29.60	29.60
Camden, S. C.	29.86	29.81	29.83	30.12	30.05	30.06	29.64	29.53	29.56
Canton, N. Y.	29.31	29.29	29.30	29.66	29.61	29.55	28.98	28.82	28.91
Carmel, Me.	29.49	29.47	29.48	29.79	29.75	29.79	28.90	29.07	29.11
Cedar Keys, Fa.(2)	29.92	29.92	29.92	30.03	30.03	30.04	29.83	29.87	29.87
Chapel Hill, N. C.	29.41	29.38	29.40	29.63	29.61	29.63	29.21	29.13	29.18
Charleston, S. C. (3)	....	....	....	....	....	....	....	....	....
Chestertown, Md. (4)	29.64	29.61	29.62	29.83	29.81	29.81	29.34	29.29	29.28
Concord, N. H.	29.53	29.50	29.51	29.82	29.81	29.82	29.06	29.17	29.12
Deaf &Dumb Inst. N. Y.	29.87	29.85	29.86	30.11	30.11	30.07	29.57	29.54	29.44
Delaware College. (5)	29.75	29.71	29.78	29.95	29.91	29.97	29.48	29.41	29.69
Detroit, Mich.	29.40	29.39	29.41	29.60	29.69	29.70	28.89	29.09	29.03
Dubuque, Iowa	29.24	29.19	29.20	29.61	29.52	29.55	28.96	28.91	28.85
Frederick City, Md.	29.49	29.45	29.47	29.74	29.71	29.66	29.20	29.09	29.16
Gardiner, Me.	29.69	29.67	29.68	29.97	29.98	29.97	29.16	29.28	29.31
Germantown, Ohio	29.13	29.11	29.12	29.39	29.39	29.38	28.76	28.87	28.75
Gettysburg, Pa.	29.36	29.32	29.34	29.58	29.61	29.58	29.02	28.92	29.00
Glenwood, Tenn.	29.51	29.47	29.50	29.77	29.72	29.73	29.23	29.30	29.26
Granville, Ohio	28.89	28.89	28.91	29.19	29.16	29.16	28.48	28.59	28.72
Great Falls, N. H.	29.52	29.48	29.51	29.81	29.82	29.82	29.04	29.12	29.05
Hamilton, Canada	29.59	....	29.59	29.84	....	29.80	29.05	....	29.35
Harrisburg, Pa.	29.66	29.63	29.64	29.93	29.90	29.97	29.30	29.24	29.29
Hillsborough, Ohio	28.82	28.80	28.80	29.11	29.08	29.07	28.40	28.50	28.47
Jackson, Ohio, (1)	28.91	28.94	28.87	29.17	29.25	29.21	28.52	28.56	28.56
Jacksonville, Fa.	30.10	30.06	30.08	30.27	30.27	30.28	29.92	29.85	29.87
Key West, Fa.	29.70	29.74	....	29.78	29.84	....	29.58	29.60	....
Knox Hill, Fa.	29.86	29.80	29.81	30.00	29.97	29.99	29.67	29.60	29.63
Knoxville, Tenn.	29.03	28.98	29.01	29.29	29.22	29.24	28.74	28.64	28.72
Lima, Pa.	29.77	29.73	29.74	30.03	30.02	30.00	29.47	29.40	29.39
Madison, Ind.	29.44	29.41	29.40	29.72	29.63	29.69	29.05	29.14	29.08
Madrid, N. Y.	29.55	29.53	29.54	29.93	29.85	29.74	29.19	29.08	29.19
Manchester, Ill.	28.62	28.62	28.61	28.96	28.93	28.90	28.44	28.31	28.30
Manchester, N. H.	29.88	29.85	29.87	30.16	30.15	30.15	29.46	29.52	29.43
Milton, Ind.	29.00	28.98	29.00	29.28	29.25	29.28	28.53	28.64	28.60
Milwaukie, Wis.	29.21	29.21	29.19	29.51	29.50	29.51	28.65	28.93	28.83
Morrisville, Pa. (1)	30.04	30.02	30.02	30.30	30.28	30.30	29.80	29.70	29.65
Muscatine, Iowa	29.39	29.37	29.38	29.73	29.65	29.68	29.18	29.10	29.05
Nantucket, Mass.	29.92	29.91	29.92	30.17	30.15	30.17	29.41	29.60	29.63
Newark, N. J,	29.92	....	29.89	30.20	....	30.50	29.57	....	29.50
New Bedford, Mass.	29.83	29.80	29.81	30.07	30.07	30.08	29.40	29.48	29.42
Newburyport, Mass.	29.89	29.83	29.84	30.18	30.08	30.17	29.38	29.50	29.38
New Orleans, La.	30.17	30.14	30.17	30.30	30.29	30.29	30.02	29.98	30.00
New Wied, Texas (1)	28.69	28.65	28.64	28.90	28.80	28.85	28.50	28.50	28.50
Norristown, Pa.	29.83	29.80	29.81	30.09	30.06	30.03	29.53	29.46	29.45
North Attleboro, Mass.	29.73	29.69	29.71	30.00	29.98	30.01	29.29	29.38	29.27
Oberlin, Ohio	29.02	29.00	29.02	29.28	29.27	29.30	28.56	28.67	28.72
Oswego, N. Y.(1)	29.52	29.53	29.53	29.76	29.76	29.74	29.13	29.18	29.18

* For changes in the hours of observation, see tables of Thermometer for this month.
(1) Not corrected for temperature.
(2) For the last 16 days of the month only.
(3) Max. for the month, 30.40 Min. 29.89.
(4) For the last 22 days of the month only.
(5) Incomplete.

NAME OF STATION.	Mean.			Maxima.			Minima.		
	7 A. M.	2 P. M.	9 P. M.	7 A. M.	2 P. M.	9 P. M.	7 A. M.	2 P. M.	9 P. M.
Oxford, Miss	29.53	29.51	29.52	29.70	29.69	29.65	29.36	29.36	29.30
Penn Yan, N. Y	29.15	29.16	29.17	29.39	29.46	29.37	28.75	28.69	28.80
Perrysburg, Ohio	29.28	29.26	29.27	29.60	29.57	29.54	28.81	28.96	28.90
Philadelphia, Pa	29.76	29.73	29.73	30.02	29.98	29.94	29.47	29.42	29.39
Pittsburg, (Oak. Sta.) Pa	28.86	28.84	28.84	29.11	29.07	29.08	28.48	28.42	28.52
Pittsburg, Penn	....	....	....	....	....	....	....	....	....
(Marine Hospital.)(1)	28.87	28.87	28.86	29.15	29.19	29.16	28.54	28.50	28.50
Pomfret, Conn	29.85	29.81	29.82	29.09	29.06	29.10	28.46	28.54	28.43
Pottsville, Pa	29.14	29.10	29.12	29.41	29.38	29.36	28.78	28.74	28.76
Princeton, Mass	28.77	28.74	28.75	29.03	29.01	29.03	28.34	28.43	28.30
Providence, R. I	29.69	29.67	29.70	29.95	29.94	29.96	29.25	29.37	29.25
Quasqueton, Iowa	28.99	28.96	28.96	29.37	29.29	29.28	28.69	28.67	28.63
Sag Harbor N. Y	29.61	29.59	29.62	29.84	29.86	29.92	29.23	29.31	29.24
Savannah, Ga	30.04	30.00	30.02	30.27	30.23	30.24	29.84	29.75	29.79
St. Johnsbury, Vt	29.09	29.07	29.11	29.35	29.30	29.77	28.70	28.70	28.70
St. Louis, Mo	29.40	29.37	29.38	29.74	29.71	29.67	29.19	29.11	29.11
St. Martins, C. E	29.76	29.75	29.75	30.10	30.15	30.09	29.46	29.27	29.38
Smithsonian Inst	29.95	29.91	29.91	30.22	30.19	30.16	29.63	29.53	29.59
Southwick, Mass	29.89	29.85	....	30.17	30.11	....	29.48	29.56	....
Sparta, Ga	29.39	28.36	29.39	29.58	29.56	29.56	29.22	29.15	29.18
Spencertown, N. Y	29.23	29.20	29.21	29.56	29.48	29.56	28.88	28.83	28.68
Springdale, Ky	29.34	29.33	29.32	29.60	29.65	29.58	28.98	29.12	29.07
Steuben, Me	29.95	29.91	29.94	30.25	30.24	30.20	29.43	29.52	29.62
Superior, Wis	29.10	29.08	29.06	29.43	29.38	29.38	28.70	28.74	28.42
The Rock, Ga. (2)	....	....	....	....	....	....	....	....	....
Upper Alton, Ill	29.39	29.36	29.38	29.72	29.65	29.65	29.20	29.12	29.12
Urbana Ohio	28.66	28.66	28.68	28.98	28.97	29.00	28.28	28.40	28.38
Warrington, Fa	29.99	29.99	29.95	30.12	30.16	30.12	29.81	29.76	29.75
Westfield, Mass	29.75	29.75	29.74	29.92	29.90	29.80	29.62	29.66	29.61
Williamstown, Mass	29.20	29.15	29.17	29.46	29.40	29.42	28.81	28.90	28.63
Worcester, Mass.	29.41	29.38	29.39	29.66	29.65	29.67	29.01	29.10	28.94
Zansville, Ohio	29.07	29.05	29.06	29.34	29.31	29.31	28.76	28.85	28.88

(1) Obs. taken at Sunrise 9 A. M. 3 P. M. and 9 P. M. Mean at 9 A. M. 28.88, Max. 29.19 Min. 28.50. (2) Max. for the month 29.80. Min. 29.45. Mean 29.62, Obs. taken noon 3 P. M. and evening.

NAME OF STATION.	Mean.			Maxima.			Minima.		
	7 A. M.	2 P. M.	9 P. M.	7 A. M.	2 P. M.	9 P. M.	7 A. M.	2 P. M.	9 P. M.
Alexandria, Va	30.02	29.97	30.00	30.28	30.26	30.21	29.81	29.80	29.83
All Saints, S. C	30.04	30.04	30.04	30.21	30.22	30.21	29.94	29.92	29.85
Amherst, Mass	29.79	29.75	29.78	30.07	30·09	30.05	29.57	29.50	29.54
Ann Arbor, Mich. (Winchell,)	29.07	29.05	29.05	29.27	29.24	29.23	28.84	28.84	28.82
Battle Creek, Mich	29.01	28.99	29.01	29.17	29.14	29.17	28.74	28.80	28.85
Bedford, Pa	29.13	29.10	29.09	29.38	29.36	29.32	28.98	28.94	28.95
Beloit, Wis	29.16	29.15	29.17	29.32	29.29	29.32	28.83	28.90	29.01
Bloomfield, N. J	29.86	29.82	29.84	30.15	30.12	30.13	29.70	29.60	29.61
Burlington, Vt	29.67	29.65	29.65	29.91	29.86	29.86	29.40	29.32	29.38
Burlington, N. J. (1)	29.99	29.98	29.98	30.23	30.23	30.20	29.80	29.81	29.80
Camden, S. C	29.92	29.87	29.90	30.10	30.08	30.09	29.74	29.71	29.74
Canton, N. Y. (2)	29.42	29.43	29.43	29.64	29.61	29.66	29.19	29.17	29.22
Carmel, Me	29.64	29.59	29.62	29.91	29.90	29.96	29.38	29.34	29.36
Cedar Keys, Fa	29.93	29.93	29.94	30.05	30.03	30.05	29.84	29.85	29.86
Chapel Hill, N. C	29.48	29.45	29.47	29.66	29.63	29.66	29.35	29.30	29.32
Charleston, S. C. (3)	....	....	....	....	....	....	....	....	....
Chestertown, Md	29.72	29.70	29.71	29.94	29.90	29.88	29.55	29.52	29.57
Concord, N. H	29.66	29.65	29.66	29.91	29.92	29.96	29.46	29.38	29.44
Deaf & Dumb Inst. N. Y.	29.94	29.95	29.96	30.25	30.18	30.20	29.80	29.80	29.80
Detroit, Mich	29.51	29.49	29.49	29.72	29.69	29.67	29.29	29.24	29.27
Dubuque, Iowa	29.29	29.26	29.27	29.47	29.39	29.46	28.99	29.03	29.07
Frederick, Md	29.59	29.56	29.58	29.78	29.76	29.75	29.45	29.36	29.42
Gardiner, Me	29.81	29.79	29.80	30.01	30.06	29.11	29.58	29.43	29.50
Germantown, O	29.23	29.20	29.22	29.43	29.40	29.36	29.07	29.01	29.08
Gettysburg, Pa	29.46	29.43	29.43	29.72	29.68	29.66	29.30	29.23	29.27
Glenwood, Tenn	29.57	29.53	29.55	29.76	29.69	29.73	29.45	29.41	29.44
Granville, Ohio	29.00	28.97	29.00	29.20	29.18	29.16	28.84	28.77	29.73
Harrisburg, Pa	29.76	29.72	29.73	30.02	29.98	29.93	29.60	29.56	59.57
Hillsborough, Ohio	28.91	28.88	28.88	29.10	29.08	29.07	28.76	28.71	28.75
Jacksonville, Fa	30.17	30.13	30.16	30.28	30.26	30.30	30.06	30.02	30.04
Jackson, Ohio,	29.36	29.39	29.36	29.67	29.71	29.70	28.87	28.85	28.87
Key West, Fa	29.76	29.81	....	29.85	29.89	....	29.70	29.72	....
Knox Hill, Fa	29.92	29.85	29.89	30.02	29.98	30.02	29.78	29.79	29.80
Lima, Pa	29.86	29.83	29.85	30.16	30.11	30.12	29.66	29.61	29.70
Manchester, Ill	28.65	28.62	28.62	28.76	28.75	28.75	28.54	28.50	28.52
Manchester, N. H	30.01	30.00	30.01	30.28	30.25	30.29	29.79	29.72	29.78
Milton, Ind	29.06	29.04	29.08	29.22	29.19	29.23	28.92	28.84	28.96
Milwaukie, Wis	29.26	29.26	29.28	29.47	29.49	29.45	28.95	29.05	29.09
Morrisville, Pa. (1)	30.11	30.11	30.12	30.40	30.40	30.35	29.95	29.90	29.90
Muscatine, Iowa	29.41	29.39	29.39	29.61	29.54	29.54	29.23	29.26	29.24
Nantucket, Mass	30.02	30.02	30.02	30.32	30·31	30.28	29.77	29.82	29.82
Newark, N. J	30.03	....	30.03	30.32	....	30.28	29.89	....	29.89
New Bedford, Mass	29.95	29.92	29.93	30.26	30.18	30.22	29.70	29.75	29.72
Newburyport, Mass	30.00	29.99	29.99	30.31	30.37	30.29	29.77	29.70	29.73
New Harmony, Ind	29.62	29.60	29.61	29.79	29.76	29.75	29.46	29.47	29.49
New Orleans, La	30.21	30.19	30.19	30.37	30.28	30.32	30.07	30.07	30.05
New Wied, Texas. (1)	28.70	28.61	28.67	28.85	28.80	28.80	28.60	28.60	28.60
Norristown, Pa	29.91	29.88	29.89	30.19	30.15	30.15	29.75	29.68	29.73
North Attleboro, Mass	29.85	29.82	29.84	30.15	30.09	30.13	29.60	29.59	29.60
Oakland Station, Pittsburg, Pa	28.97	28.94	28.96	29.15	29.16	29.14	28.80	28.75	28.80
Oberlin, Ohio	29.12	29.10	29.11	29.30	29.32	29.30	28.92	28.87	28.91
Oswego, N. Y. (1)	29.66	29.67	29.65	29.88	29.88	29.83	29.40	29.45	29.49
Oxford, Miss	29.57	29.53	29.55	29.69	29.68	29.66	29.46	29.43	29.44
Penn Yan, N. Y	29.29	29.29	29.28	29 49	29.50	29.49	29.10	29.04	29.06

* For changes in the hours of observation, see tables of Thermometer for this month.
(1) Not corrected for temperature.
(2) Incomplete.
(3) Max. for the month 30. 42, Min. 30.05.

NAME OF STATION.	Mean.			Maxima.			Minima.		
	7 A. M.	2 P. M.	9 P. M.	7 A. M.	2 P. M.	9 P. M.	7 A. M.	2 P. M.	9 P. M.
Perrysburg, Ohio	29.38	29.36	29.35	29.59	29.57	29.53	29.16	29.16	29.17
Philadelphia, Pa.	29.87	29.84	29.85	30.13	30.13	30.09	29.70	29.64	29.68
Pittsburg, Pa (1)	28.98	28.98	28.97	29.18	29.18	29.19	28.79	28.86	28.87
Pomfret, Conn	28.92	28.94	28.93	29.22	29.16	29.19	28.81	28.65	28.75
Pottsville, Pa	29.25	29.22	29.23	29.51	29.48	29.47	29.07	29.02	29.04
Princeton, Mass	28.92	28.88	28.87	29.49	29.16	29.87	28.68	28.66	28.18
Providence, R. I.	29.82	29.79	29.81	30.09	30.05	30.09	29.57	29.62	29.61
Quasqueton, Iowa	29.05	29.04	29.03	29.29	29.19	29.20	28.77	28.88	28.82
Sag Harbor, N. Y.	29.71	29.69	29.72	29.99	29.97	30.00	29.52	29.48	29.49
St. Louis, Mo	29.46	29.41	29.42	29.61	29.57	29.58	29.33	29.29	29.27
St. Johnsbury, Vt	29.22	29.19	29.20	29.50	29.38	29.39	28.99	28.96	29.00
St. Martin, C. E.	29.85	29.81	29.84	30.10	30.04	30.06	29.59	29.51	29.58
Savannah, Ga	30.12	30.07	30.10	30.27	30.24	30.26	29.96	29.93	29.95
Smithsonian Inst	30.04	30.00	30.01	30.32	30.27	30.24	29.87	29.81	29.84
Southwick, Mass	30.01	29.97	....	30.29	30.24	....	29.78	29.78	....
Sparta, Ga	29.46	29.41	29.45	29.59	29.55	29.59	29.36	29.30	29.33
Spencertown, N. Y.	29.35	29.34	29.34	29.57	29.56	29.53	29.18	29.12	29.17
Springdale, Ky.	29.41	29.39	29.38	29.63	29.55	29.56	29.28	29.25	29.27
Steuben, Me.	30.07	30.05	30.05	30.33	30.36	30.37	29.80	29.67	29.69
Superior, Wis	29.13	29.13	29.15	29.39	29.40	29.42	28.76	28.90	28.90
The Rock, Ga. (2)	....	....	....	....	....	....	....	....	....
Upper Alton, Ill	29.45	29.42	29.44	29.59	29.56	29.56	29.29	29.30	29.28
Urbana, Ohio	28.82	28.77	28.79	28.99	29.00	29.00	28.63	28.61	28.64
Warrington, Fa	30.02	30.06	30.02	30.14	30.19	30.16	29.91	29.90	29.88
Westfield, Mass	29.87	29.87	29.88	30.06	30.00	30.05	29.72	29.74	29.76
Williamstown, Mass	29.31	29.27	29.28	29.59	29.51	29.41	29.11	29.00	29.09
Worcester, Mass	29.51	29.49	29.50	29.82	29.77	29.79	29.27	29.28	29.30
Zanesville, Ohio	29.15	29.13	29.14	29.37	29.36	29.32	28.99	28.95	29.00

(1) Obs. taken at Sunrise, 9 A. M. 3 P. M. and 9 P. M. Mean at 9 A. M. 28.99, Max. 29.29, Minimum 28.99.

(2) Mean for the month 29.45, Max. 29.60, Min. 29.35, obs. taken noon, 3 P. M. and evening.

NAME OF STATION.	Mean.			Maxima.			Minima.		
	7 A. M.	2 P. M.	9 P. M.	7 A. M.	2 P. M.	9 P. M.	7 A. M.	2 P. M.	9 P. M
Alexandria, Va	30.05	30.02	30.04	30.35	30.34	30.34	29.73	29.75	29.74
All Saints, S. C	30.00	29.99	30.01	30.15	30.16	30.22	29.82	29.83	29.80
Amherst, Mass	29.84	29.82	29·80	30.15	30·10	30.11	29.33	29.39	29.23
Ann Arbor, Mich	29.12	29.11	29.11	29.47	29.37	29.39	28.75	28.86	28.75
Auburn, Ala	29.39	29.35	29.37	29.54	29.50	29.55	29.25	29.25	29.17
Battle Creek, Mich	29.08	29.06	29.07	29.36	29.34	29.33	28.79	28.82	28.76
Bedford, Pa	29.15	29.13	29.13	29.45	29.42	29.42	28.83	28.78	28.82
Beloit, Wis	29.23	29.21	29.23	29.54	29.46	29.47	28.98	28.93	28.98
Bloomfield, N. J	29.92	29.92	29.91	30.37	30.36	30.38	29.49	29.35	29.39
Burlington, Vt	29.71	29.69	29.68	30.05	30.00	29.95	29.14	29.13	29.10
Burlington, N. J.(1)	30.03	30.02	30.06	30.30	30.30	30.33	29.63	29.64	29.59
Camden, S. C	29.90	29.86	29.89	30.09	30.06	30.11	29.72	29.64	29.69
Canton, N. Y	29.47	29.42	29.45	29.83	29.80	29.71	28.95	28.85	28.90
Carmel, Me.(2)	29.76	29.72	29.73	29.93	29.87	29.93	29.49	29.47	29.46
Cedar Keys, Fa	29.89	29.90	29.89	30.00	30.03	30.02	29.79	29.81	29.80
Chapel Hill, N. C	29.47	29.45	29.46	29.69	29.68	29.68	29.26	29.25	29.26
Charleston, S. C. (3)	....	....	....	....	....	....	....	....	....
Chestertown, Md	29.75	29.73	29.74	30.00	29.99	29.98	29.41	29.30	29.34
Concord, N. H	29.71	29.68	29.69	30.00	29.98	29.98	29.15	29.09	29.03
Detroit, Mich	29.54	29.51	29.50	29.82	29.81	29.74	29.27	29.21	29.15
Deaf &Dumb Inst. N. Y	30.01	29.99	30.01	30.33	30.31	30.31	29.64	29.48	29.54
Dubuque, Iowa	29.36	29.35	29.34	29.61	29.63	29.63	29.09	29.08	29.12
Frederick City, Md	29.63	29.59	29.62	29.93	29.89	29.88	29.31	29.22	29.28
Gardiner, Me	29.83	29.79	29.80	30.10	30.07	30.10	29.21	29.31	29.21
Germantown, Ohio	29.27	29.24	29.27	29.55	29.51	29.51	29.02	29.03	29.03
Gettysburg, Pa	29.49	29.47	29.48	29.80	29.79	29.75	29.12	29.02	29.11
Glenwood, Tenn.(4)	29.58	29.54	29.56	29.80	29.73	29.71	29.40	29.35	29.27
Granville, Ohio	29.04	29.00	29.03	29.32	29.27	29.28	28.74	28.76	28.79
Harrisburg, Pa	29.79	29.75	29.77	30.10	30.10	30.07	29.44	29.33	29.39
Hillsborough, Ohio	28.93	29.91	28.92	29.19	29.16	29.17	28.69	28.69	28.71
Jacksonville, Fa	30.12	30.06	30.13	30.26	30.21	30.31	30.01	29.94	29.97
Jackson, Ohio, (1)	29.39	29.41	29.41	29.64	29.63	29.65	29.18	29.15	29.12
Key West, Fa	29.72	29.76	....	29.84	29.88	....	29.53	29.61	....
Knox Hill, Fa. (5)	29.87	29.83	29.86	29.98	29.94	29.97	29.70	29.69	29.71
Lima, Pa	29.91	29.87	29.90	30.24	30.23	30.21	29.51	29.42	29.45
Madrid, N. Y	29.81	29.81	29.81	30.34	30.43	30.23	29.43	29.41	29.41
Manchester, Ill	28.69	28.67	28.68	29.01	29.00	28.92	28.52	28.56	28.50
Manchester, N. H(6)	30.01	30.02	30.02	30.33	30.31	30.32	29.50	29.69	29·72
Milton, Ind	29.09	29.06	29.09	29.38	29.26	29.30	28.89	28.89	28.93
Milwaukie, Wis	29.33	29.32	29.31	29.57	29.61	29.57	29.06	29.07	29.03
Montreal, Ca. (7)	29.83	29.82	29.81	29.99	29.96	29.98	29.51	29.48	29.52
Morrisville, Pa. (1)	30.16	30.15	30.16	30.50	30.45	30.45	29.80	29.70	29.70
Muscatine, Iowa	29.48	29.46	29.46	29.82	29.78	29.71	29.23	29.20	29.20
Nantucket, Mass	30.05	30.04	30.05	30.33	30.34	30.37	29.55	29.56	29.47
Newark, N. J, (8)	30.10	....	30.06	....	....	....	....	....	....
New Bedford, Mass	30.07	30.06	30.07	30.27	30.28	30.27	29.45	29.41	29.37
Newburyport, Mass	30.05	30.01	20.02	30.35	30.30	30.31	29.48	29.43	29.38
New Orleans, La	30.17	30.15	30.17	30.36	30.34	30.36	29.89	29.89	29.87
New Wied, Texas (1)	28.67	28.63	28.64	28.82	28.75	28.80	28.52	28.50	28.50
Norristown, Pa	29.94	29.92	29.93	30.29	30.28	30.26	29.57	29.45	29.49
North Attleboro, Mass	29.88	29.86	29.88	30.21	30.18	30.19	29.36	29.36	29.27
Oberlin, Ohio	29.17	29.15	29.18	29.48	29.44	29.44	28.81	28.84	28.85
Oswego, N. Y.(1)	29.67	29.70	29.67	29.93	29.97	29.90	29.25	29.25	29.27
Oxford, Miss (9)	29.50	29.49	29.49	29.63	29.65	29.64	29.38	29.34	29.35
Penn Yan, N. Y	29.31	29.32	29.29	29.59	29.56	29.52	28.88	28.83	28.88
Perrysburg, Ohio	29.43	29.40	29.41	29.74	29.69	29.67	29.09	29.11	29.08

* For changes in hours of observations, see tables of Thermometer for this month.

(1) Not corrected for temperature. (2) Obs. omitted from the 5th to the 25th of the month.

(3) Max. for the month 30.38, Min. 30.02. (4) Obs. omitted from the 13th to the 23d.

(5) Obs. omitted from the 20th to the 30th of the month. (6) Incomplete.

(7) For 8 days only —11th to 18th (8) Obs. taken at 7 A. M. and 6 P. M.

(9) For the first 14 days of the month only.

NAME OF STATION.	Mean.			Maxima.			Minima.		
	7 A. M.	2 P. M.	9 P. M.	7 A. M.	2 P. M.	9 P. M.	7 A. M.	2 P. M.	9 P. M.
Philadelphia, Pa..........	29.93	29.91	29.92	30.25	30.23	30.22	29.60	29.51	29.52
Pittsburg, Pa(Oak. Sta.)	29·06	29·04	29.05	29.34	29.34	29.36	28.72	28.77	28.80
Pittsburg, Pa. (Marine Hospital,) (1)............	.... 29.01	.... 28.99	.... 29.01	.... 29.32	.... 29.29	.... 29.30	.... 28.67	.... 28.67	.... 28.68
Pomfret, Conn............	28.98	28.96	28.95	29.28	29.26	29.25	28.43	28.46	28.42
Port Gibson, Miss. (2)...	29.83	29.80	29.81	29.88	29.86	29.88	29.67	29.63	29.62
Pottsville, Pa..............	29·29	29.26	29.28	29.63	29.58	29.59	28·89	28.88	28.86
Princeton, Mass..........	28.93	28.90	28.91	29.23	29.21	29.20	28.43	28.37	28.33
St. Louis, Mo..............	29.50	29.48	29.48	29.72	29.75	29.72	29.29	29.26	29.27
Sag Harbor, N. Y........	29.72	29.70	29.74	30.06	30.06	30.08	29.30	29.21	29.19
St. Martins, C. E.........	29.86	29.82	29.85	30·20	30.12	30.12	29.30	29.20	29.20
Savannah, Ga.............	30.08	30.03	30.07	30.19	30.21	30.26	29.93	29.86	29.92
Smithsonian Inst.........	30.09	30.05	30.06	30.42	30.38	30.37	29.73	29.68	29.70
Southwick, Mass.(3)......	30.03	29.95	....	30.32	30.21	....	29.48	29.48	....
Sparta, Ga..................	29.43	29.40	29.42	29.54	29.55	29.54	29.32	29.27	29.28
Spencertown, N. Y.......	29.35	29.35	29.34	29.66	29.66	29.62	28.93	28.85	28.85
Springdale, Ky............	29.43	29.41	29.41	29.68	29.67	29.67	29.24	29.25	29.22
Steuben, Me...............	30.08	30.04	30.05	30.38	30.35	30.33	29.39	29.46	29.40
Superior, Wis..............	29.22	29.20	29 21	29.46	29.55	29.52	28.75	28.78	28.77
The Rock, Ga. (2).........	....	....	....	....	....	....	....	....	....
Upper Alton, Ill..........	29.49	29.46	29.48	29.79	29.72	29.71	29.28	29.28	29.30
Urbana, Ohio..............	28.81	28.79	28.82	29.09	29.04	29.08	28.55	28.57	28.59
Warrington, Fa...........	30.01	30.05	30.01	30.13	30·19	30.13	29.83	29.84	29.84
Westfield, Mass...........	29.91	29.90	29.90	30.14	30.11	30.10	29.72	29.69	29.71
Williamstown, Mass......	29.33	29.30	29.35	29.65	29.62	29.61	28.87	28.74	28.79
Worcester, Mass..........	29.52	29.43	29.47	29.83	29.83	29.78	29.04	28.96	28.92
Zanesville, Ohio..........	29.18	29.14	29.16	29.46	29.40	29.42	28.87	28.89	28.94

(1) Obs. taken at Sunrise, 9 A. M. 3 P. M. and 9 P. M. Mean at 9 A. M. 29.02, Max. 29.33, Min. 28.68.
(2) For the last 12 days of the month only.
(3) Incomplete.
(4) Mean for the month 29·41 Max. 29.50 Min. 29.25, obs. taken morning 3 P. M. and evening, Record incomplete.

NAME OF STATION.	Mean.			Maxima.			Minima.		
	7 A. M.	2 P. M.	9 P. M.	7 A. M.	2 P. M.	9 P. M.	7 A. M.	2 P. M.	9 P. M.
Alexandria, Va	30.13	30.08	30.09	30.35	30.29	30.31	29.91	29.78	29.76
All Saints, S. C	30.06	30.04	30.03	30.23	30.28	30.22	29.84	29.82	29.80
Amherst, Mass	29.87	29.82	29.86	30.18	30.10	30.17	29.45	29.40	29.51
Ann Arbor, Mich	29.16	29.10	29.13	29.33	29.29	29.34	28.86	28.90	28.79
Auburn, Ala	29.42	29.38	29.41	29.60	29.56	29.58	29.27	29.17	29.26
Battle Creek, Mich	29.10	29.07	29.08	29.24	29.25	29.25	28.86	28.82	28.78
Beloit, Wis	29.25	29.22	29.25	29.39	29.39	29.39	29.04	29.00	29.00
Bloomfield, N. J	29.96	29.93	29.95	30.23	30.19	30.20	29.55	29.55	29.57
Burlington, Vt	29.77	29.74	29.74	30.07	30.02	30.03	29.37	29.34	29.32
Burlington, N. J. (1)	30.11	30.06	30.07	30.32	30.32	30.28	29.80	29.80	29.70
Camden, S. C	29.96	29.90	29.92	30.16	30.12	30.10	29.75	29.68	29.70
Canton, N. Y	29.54	29.51	29.52	29.82	29.79	29.79	29.19	29.10	29.08
Carmel, Me	29.70	29.64	29.69	30.05	30.00	30.00	29.33	29.29	29·33
Cedar Keys, Fa	29.91	29.89	29.90	30.04	30.04	30.03	29.79	29.79	29.79
Chapel Hill, N. C	29.54	29.52	29.53	29.70	29.70	29.77	29.32	29.35	29.33
Charleston, S. C. ( )	....	....	....	....	....	....	....	....	....
Concord, N. H	29.75	29.72	29.74	30.05	29.99	30 02	29.31	29.33	29.38
Detroit, Mich	29.60	29.57	29.57	29.76	29.74	29.78	29.29	29.29	29.23
Dubuque, Iowa	29.37	29.32	29.34	29.55	29.50	29.52	29.13	29.05	29.04
Frederick, Md	29.71	29.65	29.68	29.90	29.88	29.88	29.50	29.34	29.36
Germantown, Ohio	29.30	29.26	29.27	29.43	29.39	29.39	29.05	29.07	28.99
Gettysburg, Pa	29.55	29.51	29.53	29.78	29.76	29.76	29.10	29.20	29.13
Glenwood, Tenn	29.61	29.55	29.57	29.75	29.72	29.74	29.44	29.35	29.33
Granville, Ohio	29.07	29.04	29.05	29.22	29.18	29.18	28.80	28.78	28.84
Harrisburg, Pa	29.87	29.84	29.86	30.07	30.07	30.07	29.59	29.51	29.49
Hiram, Ohio	28.88	28.86	28.87	29.09	29.05	29.08	28.63	28.61	28.62
Hillsborough, Pa	28.97	28.93	28.93	29.11	29.09	29.07	28.68	28.71	28.68
Jackson, Ohio. (Gilmor) (1)	29.42	29.43	29.44	29.57	29.57	29.57	29.15	29.21	29.24
Jacksonville, Fa	30.14	30.09	30.12	30.32	30.28	30 31	29.97	29.91	29.91
Key West, Fa	29.69	29.74	....	29.81	29.84	....	29.60	29.61	....
Knox Hill, Fa	29.87	29.83	29.85	30.01	29.98	30.03	29.72	29.67	29.70
Lima, Pa	29.98	29.93	29.94	30.21	30.20	30.20	29.64	29.65	29.57
Manchester, Ill	28.69	28.69	28.67	28.87	28.84	28.80	28.56	28.53	28.53
Manchester, N. H	30.08	30.03	30.06	30.35	30.31	30.31	29.63	29.64	29.72
Milton, Ind	29.10	29.09	29.10	29.26	29.23	29.26	28.89	28.90	28.83
Milwaukie, Wis	29.37	29.34	29.34	29.53	29.54	29.52	29.18	29.08	29.07
Montreal, Ca	29.92	29.90	29.91	30.24	30.20	30.22	29.50	29.48	29.45
Morrisville, Pa. (1)	30.22	30.19	30.19	30.45	30.45	30.43	29.90	29.90	29.85
Muscatine, Iowa	29.50	29.47	29.48	29.69	29.60	29.64	29.31	29.27	29.23
N. York Deaf & Dumb Inst	30.08	30.03	30.05	30.34	30.28	30.30	29.71	29.67	29.74
Newark Del. Coll. (3)	29.91	29.86	29.90	30.15	30.08	30.10	29.61	29.59	29.78
Nantucket, Mass	30.10	30.08	30.08	30.35	30.33	30.32	29.71	29.69	29.79
Newark, N. J	30.13	....	30.08	30.39	....	30.34	29.73	....	29.78
New Bedford, Mass	30.02	29.98	30.00	30.29	30.25	30.25	29.57	29.53	29.71
Newburyport, Mass	30.10	30.09	30.07	30.39	30.36	30.36	29.66	29.63	29.77
New Orleans, La	30.20	30.17	30.20	30.33	30.29	30.34	29.93	30.04	30.01
New Wied, Texas, (1)	28.71	28.64	28.71	28.80	28.72	28.80	28.58	28.60	28.60
Norristown, Pa	30.05	30.01	30.02	30.28	30.23	30.24	29.77	29.68	29.66
North Attleboro, Mass	29.93	29.89	29.92	30.22	30.16	30.17	29.47	29.44	29.63
Oberlin, Ohio	29.20	29.17	29.17	29.35	29.33	29.33	28.86	28.90	28.92
Oswego, N. Y	29.62	29.62	29·60	29.87	29.88	29.91	29.31	29.24	29.22
Oxford, Miss. (4)	29.57	29.53	29.55	29.64	29.62	29.63	29.44	29.33	29.37
Pittsburgh, Pa. Oakland Station	29.06	29.03	29.02	29.24	29.20	29.20	28.81	28.72	28.81
Penn Yan, N. Y	29.36	29.36	29.35	29.58	29.58	29.59	29.07	28.94	28.99

* For changes in the hours of observation, see tables of Thermometer for this month.
( ) Not corrected for temperature.
(2) Maximum for the month 30·50, Min. 30·08.
(3) Record incomplete.
(4) For the last 22 days of the month only.

NAME OF STATION.	Mean.			Maxima.			Minima.		
	7 A. M	2 P. M.	9 P. M.	7 A. M.	2 P. M.	9 P. M.	7 A. M.	2 P. M.	9 P. M.
Perrysburg, Ohio	29.46	29.44	29.43	29.65	29.65	29.60	29.16	29.18	29.12
Philadelphia, Pa	30.01	29 97	29.97	30.22	30.19	30.18	29.71	29.62	29.68
Pittsburg, Pa. (Marine Hospital.) (1)	29.05	29.03	29.03	29.23	29.21	29.22	28.79	28.68	28.78
Pomfret, Conn	29.04	28.99	28.98	29.28	29.24	29.24	28.63	28.61	28.74
Pottsville, Pa	29.36	29.31	29.34	29.60	29.55	29.57	29.09	28.99	29.12
Princeton, Mass	28.96	28.93	28.96	29.24	29.19	29.27	28.55	28.46	28.63
Providence, R. I	29.90	29.86	29.90	30.17	30.14	30.17	29.44	29.42	29.59
Sag Harbor, N. Y	29.83	29.80	29.83	30.11	30.03	30.09	29.38	29.57	29.53
St. Martins, Ca. E	29.88	29.81	29.81	30.16	30.15	30.16	29.44	29.38	29.33
Savannah, Ga	30.11	30.06	30.09	30.30	30.28	30.29	29.90	29.84	29.85
St. Louis, Mo	29.51	29.45	29.47	29.65	29.57	29.60	29.34	29.30	29.32
Smithsonian Inst	30.13	30.11	30.11	30.37	30.36	30.37	29.92	29.77	29.75
Southwick, Mass	30.10	30.05	....	30.36	30.29	....	29.68	29.81	....
Sparta, Ga	29.47	29.42	29.44	29.61	29.55	29.59	29.31	29.23	29.28
Spencertown, N. Y	29.47	29.43	29.44	29.83	29.75	29.78	29.19	29.04	29.17
Springdale, Ky	29.47	29.44	29.42	29.64	.29.61	29.58	29.26	29.24	29.23
Steuben, Me	30.12	30.09	30.09	30.42	30.42	30.42	29.76	29.26	29.68
Superior, Wis	29.33	29.32	29.32	29.70	29.70	29.70	29.05	28.89	28.90
The Rock, Ga. (2)	....	....	....	....	....	....	....	....	....
Upper Alton, Ill	29.50	29.45	29.48	29.62	29.60	29.64	29.34	29.25	29.31
Urbana, Ohio	28.84	28.82	28.83	28.98	28.93	28.98	28.57	28.58	28.61
Warrington, Fa	30.05	30.06	30.02	30.18	30.20	30.18	29.94	29.81	29.86
Westfield, Mass	29.94	29.93	29.94	30.16	30.07	30.12	29.81	29.74	29.82
Williamstown, Mass	29.40	29.35	29.37	29.67	29.60	29.63	28.99	29.02	28.99
Worcester, Mass	29.56	29.53	29.53	29.81	29.88	29.85	29.16	29.20	29.23
Zanesville, Ohio	29.21	29.17	29.19	29.38	29.31	29.31	28.91	28.92	28.97

(1) Obs. taken at Sunrise, 9 A. M. 3 P. M. and 9 P. M. Mean at 9 A. M. 29.07, Max. 29.21 Minimum 28.78.

(2) Mean for the month 29.47, Max. 29.60, Min. 29.30, obs. taken morning, 3 P. M. & evening.

NAME OF STATION.	Mean.			Maxima.			Minima.		
	7 A. M.	2 P. M.	9 P. M.	7 A. M.	2 P. M.	9 P. M.	7 A. M.	2 P. M.	9 P. M.
Alexandria, Va	30.00	29.95	30.01	30.30	30.25	30.23	29.63	29.64	29.66
All Saints, S. C.	29.98	29.96	29.98	30.25	30.24	30.19	29.67	29.61	29.64
Amherst, Mass.	29.68	29.65	29.66	30.12	30.09	30.06	29.23	29.22	29.30
Ann Arbor, Mich. (Winchell,)	29.03	28.98	29.02	29.33	29.20	29.34	28.75	28.72	28.55
Auburn, Ala	29.39	29.34	29.37	29.58	29.53	29.53	29.22	29.16	29.15
Battle Creek, Mich	29.02	28.99	29.03	29.32	29.28	29.84	28.09	28.70	28.67
Bedford, Pa.	29.15	29.13	29.11	29.40	29.34	29.30	28.96	28.89	28.89
Beloit, Wis.	29.17	29.18	29.18	29.45	29.43	29.45	28.85	28.84	28.80
Bloomfield, N. J.	29.75	29.75	29.78	30.22	30.20	30.12	29.27	29.36	29.43
Burlington, Vt.	29.57	29.55	29.57	30.03	29.96	29.90	29.07	29.14	29.22
Burlington, N. J.(1)	29.93	29.91	29.93	30.20	30.24	30.20	29.50	29.52	29.60
Cincinnati Woodw. H. S. Ohio	29.42	29.41	29.41	29.65	29.62	29.62	29.18	29.14	29.17
Camden, S. C.	29.89	29.82	29.84	30.19	30.06	30.08	29.58	29.48	29.57
Camden, Ark.	29.13	29.09	29.11	29.44	29.39	29.40	28.75	28.81	28.77
Carmel, Me.	29.56	29.56	29.56	29.84	29.79	29.97	29.22	29.08	29.19
Canton, N.Y.	29.34	29.32	29.34	29.76	29.79	29.67	28.86	28.97	29.00
Cedar Keys, Fa.	29.89	29.88	29.88	30.09	30.06	30.06	29.74	29.74	29.69
Chapel Hill, N. C.	29.45	29.42	29.44	29.67	29.65	29.63	29.14	29.14	29.12
Charleston, S. C. (2)	....	....	....	....	....	....	....	....	....
Chestertown, Md.	29.77	29.73	29.76	30.01	29.96	29.93	29.32	29.33	29.33
Concord, N. H.	29.56	29.54	29.56	30.00	29.96	29.96	29.12	29.06	29.10
Detroit, Mich.	29.45	29.44	29.44	29.77	29.78	29.79	29.20	29.08	28.98
Dubuque, Iowa	29.35	29.28	29.32	29.65	29.60	29.62	28.97	28.80	28.84
Frederick, Md.	29.57	29.52	29.57	29.92	29.78	29.77	29.21	29.20	29.25
Germantown, O.	29.23	29.19	29.21	29.44	29.38	29.44	28.95	28.91	28.94
Gettysburg, Pa.	29.40	29.36	29.39	29.79	29.70	29.68	28.98	29.00	29.06
Glenwood, Tenn.	29.60	29.56	29.58	29.82	29.77	29.78	29.34	29.26	29.32
Granville, Ohio	29.00	28.96	28.96	29.19	29.15	29.15	28.73	28.64	28.68
Harrisburg, Pa.	29.69	29.68	29.72	29.97	30.03	29.98	29.25	29.33	29.38
Hillsborough, Ohio	28.86	28.84	28.85	29.10	29.04	29.05	28.63	28.57	28.58
Hiram, Ohio	28.80	28.79	28.82	29.06	29.07	29.08	28.57	28.54	28.57
Jackson, Ohio,(1)	29.28	29.32	29.30	29.48	29.58	29.48	29.08	29.06	29.07
Jacksonville, Fa.	30.12	30.06	30.09	30.36	30.29	30.32	29.91	29.80	29.83
Key West, Fa.	29.67	29.71	....	29.74	29.80	....	29.56	29.57	....
Knox Hill, Fa.	29.87	29.81	29.86	30.07	29.99	30.05	29.73	29.66	29.67
Lima, Pa.	29.80	29.79	29.80	30.18	30.12	30.08	29.33	29.41	28.40
Manchester, Ill.	28.78	28.76	28.72	29.08	29.04	29.03	28.41	28.41	28.42
Manchester, N. H.	29.84	29.91	29.81	30.15	30.11	30.18	29.48	29.47	29.43
Millersburg, Ky.	29.16	29.22	29.20	29.40	29.53	29.43	28.43	28.95	29.95
Milton, Ind.	29.03	29.03	29.03	29.23	29.20	29.33	28.73	28.75	28.76
Milwaukie, Wis.	29.30	29.26	29.29	29.91	29.58	29.53	28.91	28.93	28.91
Montreal, Ca.	29.73	29.74	29.71	30.17	30.16	30.08	29.38	29.32	29.30
Morrisville, Pa. (1)	30.05	30.02	30.04	30.45	30.40	30.35	29.60	29.65	29.70
Muscatine, Iowa	29.52	29.46	29.49	29.86	29.86	29.86	29.17	29.07	29.05
N.York Deaf &DumbInst	29.92	29.87	29.92	30.30	30.23	30.20	29.50	29.52	29.58
Newark Del. Coll.	29.74	29.69	29.74	30.07	30.05	30.07	29.34	29.39	29.36
Nantucket, Mass.	29.94	29.90	29.93	30.35	30.34	30.32	29.52	29.40	29.60
Newark, N. J.	29.97	....	29.94	30.19	....	30.19	29.48	....	29.62
New Bedford, Mass.	29.84	29.80	29.84	30.29	30.26	30.21	29.40	29.32	29.49
Newburyport, Mass.	29.90	29.87	29.94	30.35	30.35	30.34	29.45	29.48	29.46
New Orleans, La	30.27	30.25	30.27	30.44	30.38	30.40	29.91	29.92	30.02
New Wied, Texas.(1)	28.95	28.84	28.87	29.37	29.25	29.21	28.56	28.48	28.45
Norristown, Pa.	29.90	29.88	29.90	30.26	30.19	30.14	29.46	29.51	29.60
North Attleboro, Mass.	29.75	29.71	29.75	30.18	30.16	30.17	29.28	29.24	29.39

* For changes in the hours of observation, see tables of thermometer for this month.
(1) Not corrected for temperature.
(2) Max. for the month 30.37, Min. 29.82.

NAME OF STATION.	Mean.			Maxima.			Minima.		
	7 A. M.	2 P. M.	9 P. M.	7 A. M.	2 P. M.	9 P. M.	7 A. M.	2 P. M.	9 P. M.
Oberlin, Ohio	29.09	29.07	29.08	29.34	29.04	29.37	28.79	28.74	28.68
Oswego, N. Y.	29.22	29.21	29.21	29.60	29.54	29.52	28.72	28.76	28.92
Oxford, Miss	29.59	29.55	29.57	29.79	29.74	29.72	29.28	29.28	29.31
Penn Yan, N. Y.	29.17	29.15	29.18	29.96	29.42	29.48	28.65	28.79	28.92
Perrysburg, Ohio	29.36	29.33	29.33	29.74	29.70	29.72	29.07	29.00	28.94
Philadelphia, Pa	29.85	29.80	29.84	30.18	30.16	30.12	29.45	29.47	29.48
Pittsburg, Pa. (Oakland Station.)	28.91	28.88	28.90	29.15	29.12	29.13	28.58	28.60	28.62
Pittsburg, Pa. (Marine Hospital.) (1)	28.93	28.92	28.94	29.18	29.19	29.15	28.62	28.63	28.71
Pomfret, Conn	28.85	28.80	28.84	29.22	29.22	29.21	28.43	28.33	28.52
Portland, Me. (2)	29.78	29.86	29.83	30.50	30.80	30.10	29.40	29.34	29.40
Pottsville, Pa	29.17	29.14	29.17	29.56	29.48	29.46	28.78	28.80	28.83
Princeton, Mass	28.76	28.74	28.76	29.19	29.18	29.16	28.28	28.28	28.38
Providence, R. I.	29.71	29.68	29.72	30.16	30.15	30.11	29.16	29.17	29.37
Quasqueton, Iowa	29.11	29.05	29.07	29.45	29.42	29.39	28.81	28.61	28.61
St. Martins, C. E	29.68	29.66	29.68	30.15	30.10	30.03	29.27	29.23	29.19
Sag Harbor, N.Y	29.64	29.59	29.65	30.06	30.06	30.07	29.14	29.17	29.27
St. Louis, Mo	29.52	29.47	29.47	29.83	28.78	29.77	29.13	29.07	29.15
Savannah, Ga	30.06	30.00	30.04	30.31	30.24	30.29	29.81	29.70	29.76
Smithsonian Inst	30.01	29.96	30.02	30.83	30.28	30.25	29.61	29.60	29.61
Southwick, Mass. (3)	29.86	29.77	29.89	30.17	30.10	30.00	29.43	29.41	29.81
Sparta, Ga	29.44	29.39	29.43	29.64	29.53	29.61	29.25	29.19	29.22
Spencertown, N. Y.	29.37	29.33	29.36	29.81	29.73	29.71	28.81	29.00	29.05
Springdale, Ky	29.43	29.39	29.41	28.70	29.59	29.61	29.18	29.12	29.14
Steuben, Me	29.98	29.96	29.99	30.37	30.40	30.42	29.57	29·44	29.57
Superior, Wis	29.31	29.28	29.30	29.75	29.68	29.76	28.85	28.79	28.90
The Rock, Ga. (4)	....	....	....	....	....	....	....	....	....
Upper Alton, Ill	29.50	29.44	29.46	29.80	29.76	29.73	29.13	29.11	29.15
Urbana, Ohio	28.74	28.72	28.76	28.98	28.93	29.90	28.50	28.48	28.48
Warrington, Fa	30.04	30.05	30.02	30.27	30.26	30.24	29.85	29.85	29.81
Westfield, Mass	29.84	29.86	29.85	30.14	30.06	30.03	29.65	29.67	29.75
Williamstown, Mass	29.18	29.16	29.19	29.62	29.56	29.52	28.65	28.75	28.86
Worcester, Mass	29.41	29.36	29.40	29.75	29.78	29.76	28.88	28.96	29.03
Zanesville, Ohio	29.12	29.09	29.12	29.35	29.31	29.36	28.87	28·82	28.85

(1) Obs. taken at Sunrise, 9 A. M. 3 P. M. and 9 P. M. Mean at 9 A. M. 28.95, Max. 29.23, Min. 28.61,
(2) Not corrected for temperature.
(3) Incomplete.
(4) Mean for the month 29.42, Max. 29.55, Min. 29·30, obs. taken morn. 3 P. M. and evening.

NAME OF STATION.	Mean.			Maxima.			Minima.		
	7 A. M.	2 P. M.	9 P. M.	7 A. M.	2 P. M.	9 P. M.	7 A. M.	2 P. M.	9 P. M.
Alexandria, Va	30.13	30.08	30.12	30.40	30.43	30.41	29.60	29.47	29.51
All Saints, S. C	30.05	30.03	30.05	30.30	30.29	30.32	29.71	29.52	29.50
Amherst, Mass	29.84	29 80	29.80	30.22	30.13	30.12	29.16	29.13	29.15
Ann Arbor, Mich	29.10	29 05	29.08	29.49	29.36	29.41	28.56	28.61	28.60
Annapolis, Md.[1]	30.14	30.09	30.13	30.43	30.38	30.42	29.60	29.45	29.48
Auburn, Ala	29.38	29.35	29.37	29.57	29.57	29.61	29.15	29.00	29.13
Battle Creek, Mich	29.07	29.04	29.07	29.45	29.39	29.45	28.59	28.63	28.65
Bedford, Pa	29.19	29.14	29.15	29.48	29.46	29 45	28.68	28.63	28.63
Beloit, Wis	29.20	29.19	29.20	29.62	29.47	29.63	28.73	28.60	28.73
Blackwell's Island, N. Y.	30.12	30.07	30.09	30.42	30.36	30.38	29.49	29.42	29.41
Bloomfield, N. J	29.95	29.91	29.92	30.26	30.29	30.23	29.33	29.26	29.26
Burlington, Vt	29.75	29.70	29.69	30.09	30.06	30.13	29.01	28.97	29.08
Burlington N. J. (2)	30.07	30.07	30.07	30.30	30.35	30.35	29.50	29.52	29.50
Camden, Ark	59.35	29.30	29.29	29.84	29.76	29.79	28.75	28.89	28.66
Camden, S. C	29.92	29.94	29.94	30.20	30.17	30.23	29.60	29.41	29.45
Canton, N. Y	29·47	29.43	29.45	29.83	29.88	29.89	28.75	28.70	28.83
Carmel, Me	29.70	29.68	29.69	30.15	30.18	30.19	29.00	28.98	28.97
Cedar Keys, Fa	29.89	29.88	29.89	30.07	30.03	30.05	29·58	29.59	29.64
Chapel Hill, N. C	29.54	29.51	29.53	29.76	29.74	29.77	29.13	29.04	29.00
Charleston, S. C. (3)	....	....	....	....	....	....	....	....	....
Chestertown, Md	29.88	29.88	29.88	30.09	30.06	30.07	29.35	29.23	29.23
Cincinnati Woodward H.	....	....	....	....	....	....	....	....	....
School, O	29.51	29.50	29.50	29.90	29.90	29.86	29.05	29.05	29.15
Concord, N. H	29.69	29.69	29.68	30.11	30.05	30.06	29.03	29.00	29.00
Deaf & Dumb Inst. N. Y.	30.06	30.01	30.03	30.35	30.29	30.30	29.45	29.43	29.39
Detroit, Mich	29.56	29.53	29.55	29.95	29.91	29.88	29.01	29.08	29.04
Delaware College [4]	29.85	29.77	29.78	30.13	30.11	30.10	29.31	29.25	29.26
Dubuque, Iowa	29.32	29.28	29.31	29.78	29.68	29.78	28.78	28.58	28.74
Frederick, Md	29.69	29.64	29.66	29.93	29.90	29.90	29.18	29.06	29.13
Germantown, Ohio	29.32	29.29	29.32	29.71	29.67	29.63	28.85	28.83	28.97
Gettysburg, Pa	29.56	29.50	29.52	29.83	29.82	29.80	29.01	28.86	28.90
Glennwood, Tenn	29.61	29.57	29.59	29.99	29.87	29.96	29.26	29.24	29.37
Granville, Ohio	29.02	28.99	29.02	29.40	29.32	29.29	28·52	28.64	28.61
Harrisburg, Pa	29.85	29.84	29.85	30.12	30.11	30.10	29.31	29.47	29.42
Hillsborough, Ohio	28.91	28.88	28.91	29.28	29.21	29.18	28.43	28.40	28.53
Jacksonville, Fa	30.12	30.04	30.10	30.29	30.31	30.31	29.79	29.61	29.87
Jackson, Ohio.[2]	29.35	29.34	29.33	29.66	29.69	29.62	28.87	28.78	28.92
Key West, Fa.	29.64	29.67	....	29.74	29.80	....	29.49	29.50	....
Knox Hill, Fa	29.85	29.81	29.85	30.03	30.01	30.04	29.66	29.56	29.63
Lima, Pa	29.96	29.91	29.92	30.22	30.20	30.23	29.37	29.28	29.32
Manchester, Ill. [5]	28.82	28.75	28.77	29.28	29.16	29.21	28.32	28.32	28.40
Manchester, N. H	30.09	30.11	30.10	30.48	30.44	30.46	29.47	29.61	29.44
Millersburg, Ky. [5]	29.21	29.19	29.23	29.52	29.52	29.52	28.83	28.89	29.03
Milton, Ind	29.06	29.04	29.07	29.40	29.34	29.37	28.58	28.53	28.73
Milwaukie, Wis	29.30	29.30	29.31	29.71	29.71	29.70	28.82	28.91	28.90
Montreal, Ca	29.87	29.87	29.89	30.29	30.23	30.28	29.12	29.11	29.33
Morrisville, Pa. [2]	30.18	30.15	30.12	30.42	30.42	30.48	29.60	29.55	29.52
Muscatine, Iowa	29.50	29.46	29.48	29.93	29.87	29.89	28.94	28.94	29.08
Nantucket, Mass	30.03	29.99	30.03	30.40	30.35	30.34	29.37	29.43	29.40
Newark, N. J	30.10	....	30.01	30.37	....	30.35	29.54	....	29.40
New Bedford, Mass	29.96	29.94	29.95	30.32	30.28	30.23	29.25	29.27	29.28
Newburyport, Mass	30.07	30.02	30.04	30.46	30.36	30.42	29.34	29.44	29.35
New Harmony, Ind	29.70	29.67	29.67	30.13	30.00	30.01	29.34	29.29	29.39
New Orleans, La	30.21	30.18	30.22	30.44	30.36	30.43	29.95	29.90	29.95
New Wied, Texas.[2]	28.94	28.86	28.88	29.27	29.20	29.20	28.51	28.50	28.48
Norristown, Pa	30.00	29.94	29.97	30.28	30.28	30.28	29.44	29.33	29.12

(1) Obs. taken at Sunrise, 7½ A. M. 2 P. M. and 9 P. M. Mean at Sunrise 30.12 Max. 30.34, Min. 29.51.
(2) Not corrected for temperature.
(3) Max. for the month 30.39 Min. 29.68.
(4) Incomplete.
[5] See notes on tables of thermometer for this month.

NAME OF STATION.	Mean.			Maxima.			Minima.		
	7 A.M.	2 P.M.	9 P.M.	7 A.M.	2 P.M.	9 P.M.	7 A.M.	2 P.M.	9 P.M.
North Attleboro, Mass...	29.88	29.85	29.87	30.27	30.19	30.21	29.17	29.20	29.20
Oberlin, Ohio..............	29.15	29.13	29.15	29.50	29.46	29.46	29.48	28.63	28.70
Ogdensburg, N. Y.........	29.78	29.71	....	30.13	30.14	....	28.96	29.06	....
Oswego, N. Y.............	29.35	29.32	29.30	29.72	29.73	29.72	28.91	28.87	28.80
Ovid, Seneca Coll. Inst. N. Y......................	28.92	28.88	28.89	29.28	29.27	29.25	28.30	28.27	28.35
Oxford, Miss..............	29.58	29.54	29.57	29.92	29.81	29.87	29.35	29.32	29.31
Penn Yan, N. Y. (1)......	29.28	29.28	29.27	29.60	29.65	29.63	28.63	28.70	28.70
Perrysburg, Ohio (1).....	29.41	29.40	29.40	29.80	29.73	29.72	28.94	28.90	28.98
Philadelphia, Pa..........	29.27	29.93	29.94	30.22	30.20	30.21	29.43	29.36	29.33
Pittsburg, Pa., (Oakland Station)..................	29.04	28.96	29.02	29.35	29.29	29.25	28.50	28.45	28.51
Pittsburg, Pa., (Marine Hospital) (2)............	29.02	28.99	29.01	29.30	29.35	29.29	28.56	28.46	28.47
Pomfret, Conn.............	28.96	28.95	28.95	29.32	29.24	29.25	28.32	28.35	28.34
Portland, Me. (3).........	29.94	29.95	29.95	30.40	30.40	30.50	29.25	29.30	29.30
Pottsville, Pa..............	29.33	29.27	29.28	29.62	29.59	29.58	28.75	28.67	29.69
Princeton, Mass......	28.88	28.87	28.87	29.27	29.20	29.24	28.21	28.22	28.21
Quasqueton, Iowa, (4)...	29.11	29.06	29.03	29.50	29.44	29.50	28.67	28.65	28.59
St. Augustine, Fla........	30.16	30.14	30.15	30.37	30.34	30.33	29.79	29.70	29.81
St. Louis, Mo........	29.51	29.47	29.48	29.92	29.89	29.92	29.04	28.99	29.13
St. Martins, C. E..........	29.85	29.82	29.84	30.21	30.23	30.26	29.02	29.02	29.51
Savannah, Ga.............	30.10	30.05	30.09	30.29	30.30	30.35	29.76	29.61	29.68
Smithsonian Inst..........	30.15	30.09	30.13	30.44	30.38	30.41	29.61	29.46	29.49
Southwick, Mass. (4).....	30.05	30.07	30.17	30.38	30.34	30.24	29.39	29.57	30.09
Sparta, Ga..................	29.47	29.42	29.46	29.66	29.64	29.66	29.25	25.05	29.14
Spencertown, N. Y........	29.52	29.46	29.49	29.84	29.80	29.85	28.87	28.82	28.79
Springdale, Ky............	29.44	29.43	26.44	29.88	29.79	29.76	29.09	29.00	29.08
Steuben, Me................	30.09	30.06	30.05	30.59	30.58	30.53	29.23	29.37	29.40
Superior, Wis...............	29.30	29.26	29.28	29.77	29.71	29.65	28.77	28.63	28.58
The Rock, Ga. (5).........	....	....	....	....	....	....	....	....	....
Upper Alton, Ill............	29.50	29.46	29.50	29.89	29.89	29.93	29.06	29.01	29.16
Warrington, Fla. (4).....	30.00	30.01	29.99	30.21	30.17	30.19	29.81	29.80	29.79
Westfield, Mass............	29.91	29.90	29.92	30.20	30.12	30.16	29.62	29.60	29.64
Williamstown, Mass......	29.33	29.29	29.31	29.68	29.63	29.67	28.56	28.63	28.78
Woodward H. School, O.	29.51	29.50	29.50	29.90	29.90	29.86	29.05	29.05	29.15
Worcester, Mass...........	29.51	29.51	29.52	29.88	29.84	29.84	28.44	28.92	28.81
Zanesville, Ohio...........	29.21	29.16	29.20	29.58	29.50	29.48	28.91	28.70	28.78

(1) See notes on tables of Thermometer for this month.
(2) Observations taken at Sunrise, 9 A. M., 3 P. M., and 9 P. M. Mean at 9 A. M., 29.03; Max. 29.39; Min. 28.56.
(3) Not corrected for temperature.
(4) Incomplete.
(5) Mean for the month, 29.40; Max. 29.60; Min. 29.30. Observations taken morning, 3 P. M., and evening.

NAME OF STATION.	Mean.			Maxima.			Minima.		
	7 A. M.	2 P. M.	9 P. M.	7 A. M.	2 P. M.	9 P. M.	7 A. M.	2 P. M.	9 P. M.
Alexandria, Va	30.11	30.07	30.10	30.68	30.64	30.58	29.46	29.15	29.28
All Saints, S. C	30.09	30.05	30.07	30.55	30.57	30.52	29.67	29.45	29.54
Amherst, Mass	29.82	29.80	29.83	30.38	30.37	30.36	28.85	28.96	28.64
Ann Arbor, Mich. (1)	29.00	28.99	29.00	29.49	29.37	29.45	27.97	27.96	28.31
Auburn, Ala	29.42	29.38	29.41	29.75	29.73	29.77	29.02	29.02	29.16
Battle Creek, Mich	29.08	29.04	29.05	29.62	29.46	29.44	28.02	28.01	28.25
Beloit, Wis	29.20	29.17	29.21	29.69	29.62	29.61	28.29	28.59	28.29
Bloomfield, N. J. (2)	29.92	29.87	29.90	30.39	30.40	30.40	29.15	29.04	28.89
Burlington, Vt	29.66	29.64	29.66	30.20	30.14	30.21	28.61	28.71	28.56
Burlington N. J. (3)	30.04	30.06	30.01	30.50	30.54	30.54	29.30	29.20	29.10
Camden, S. C	29.99	29.94	29.97	30.47	30.44	30.42	29.51	29.33	29.47
Canton, N. Y	29.37	29.35	29.39	29.87	29.90	29.94	28.33	28.51	28.33
Cincinnati Woodw. H. S.	29.34	29.30	29.32	29.84	29.81	29.77	28.56	28.53	28.68
Carmel, Me	29.65	29.65	29.64	30.19	30.11	30.10	28.62	28.75	28.69
Cedar Keys, Fa	29.94	29.92	29.93	30.30	30.28	30.29	28.91	28.95	28.95
Chapel Hill, N. C	29.54	29.51	29.52	30.07	30.04	29.95	29.08	28.83	28.95
Charleston, S. C. (4)	....	....	....	....	....	....	....	....	....
Chestertown, Md	29·87	29.81	29.83	30.35	30.33	30.29	29.22	28.90	28.96
Cincinnati Ohio, (5)	....	....	....	....	....	....	....	....	....
Concord, N. H	29.66	29.64	29.65	30.17	30.13	30.16	28.68	28.88	28.53
Carlisle Dickinson Coll. Pa	29.29	29.21	29.24	29.69	29.75	29.71	28.59	28.16	28.30
Detroit, Mich	29.52	29.49	29.52	30.03	30.04	29.90	28.47	28.41	28.73
Dubuque, Iowa	29.38	29.33	29.34	29.94	29.81	29.82	28.54	28.53	28.25
Frederick, Md	29.70	29.64	29.67	30.35	30.29	30.19	29.01	28.61	28.80
Germantown, Ohio	29.32	29.28	29.32	29.86	29.77	29.83	28.32	28.67	28.71
Gettysburg, Pa	29.51	29.47	29.49	30.06	30·01	29.94	28.84	28.47	28.62
Glennwood, Tenn	29.64	29.60	29.62	30.02	29.96	30.01	29.04	29.19	29.27
Granville, Ohio	28.98	28.98	28.96	29.44	29.40	29.45	27.99	28.10	28.02
Harrisburg, Pa	29.82	29.78	29.82	30.35	30.32	30.31	29.11	28.83	28.87
Hillsborough, Ohio	28.89	28.87	28.89	29.41	29.33	29.40	27.91	28.12	28.36
Jacksonville, Fa	30.24	30.15	30.21	30.56	30.58	30.57	29.96	29.81	29.83
Key West, Fa	29.74	29.76	....	29.94	29.96	....	29.60	29.61	....
Lima, Pa	29.92	29.89	29.91	30.48	30.44	30.42	29.27	29.01	28.92
Manchester, Ill. (2)	28.87	28.79	28.80	29.31	29.18	29.21	28.22	28.39	27.91
Manchester, N. H	29.93	29.91	29.94	30.34	30.32	30.33	29.08	29.18	29.34
Millersburg, Ky	29.21	29.23	29.24	29.52	29.54	29.48	28.33	28.49	28.73
Milton, Ind	29.05	29.03	29.09	29.51	29.42	29.42	28.08	28.28	28.37
Milwaukie, Wis	29.27	29.26	29.30	29·77	29.70	29.68	28.25	28.60	28.43
Montreal, Ça	29.83	29.79	29.82	30.33	30.34	30.38	28.71	28.91	28.78
Morrisville, Pa. (3)	30.15	30.13	30.15	30.65	30.65	30.65	29.45	29.35	29.20
Muscatine, Iowa	29.56	29.49	29.53	30.04	29.98	29.97	28.77	28.86	28.56
N. York Deaf & Dumb Inst	30.02	29.99	30.01	30.47	30.46	30.49	29.25	29.15	29.08
Nantucket, Mass	30.01	29.96	29.97	30.54	30.47	30.49	29.26	29.37	29.11
Newark, N. J. (1)	30.06	....	30.03	....	....	....	....	....	....
New Bedford, Mass	29.93	29.90	29.91	30.47	30.39	30.39	29.08	29.22	28.89
Newburyport, Mass	30.02	30.00	30.02	30.55	30.49	30.53	29.22	29.20	28.92
New Harmony, Ind	29.73	29.69	29.72	30.20	30.16	30.22	29.02	29.22	28.91
New Orleans, La	30.28	30.25	30.28	30.63	30.61	30.66	29.96	29.93	30.00
New Wied, Texas. (3)	29.10	29.00	29.01	29.55	29.49	29.50	28.70	28.68	28.64
Norristown, Pa	30.04	30.01	30.04	30.55	30.52	30.48	29.27	29.39	29.45
North Attleboro, Mass.	29.83	29.81	29.84	30.37	30.32	30.36	28.96	29.08	28.76
Ovid Seneca Coll. Inst. N. Y	28.85	28.83	28.84	29.29	29.27	29.26	28.07	27.89	27.91
Oberlin, Ohio	29.11	29.10	29.10	29.70	29.57	29.53	28.02	28.10	28.31
Oswego, N. Y	29.29	29.29	29.29	29.80	29.81	29.85	28.41	28.24	28.28

(1) Incomplete.
(2) See notes on tables of thermometer for this month.
(3) Not corrected for temperature.
(4) Maximum 30.71, Min. 29.71.
(5) Max. for the month 30.15 Min. 28.38.

NAME OF STATION.	Mean.			Maxima.			Minima.		
	7 A. M.	2 P. M.	9 P. M.	7 A. M.	2 P.M.	9 P. M.	7 A. M.	2 P. M.	9 P. M.
Oxford, Miss.	29.65	29.62	29.63	30.01	30.41	30.12	20.19	29.19	29.13
Penn Yan, N. Y	29.20	29.23	29.22	29.65	29.74	29.70	28.43	28.31	20.18
Philadelphia, Pa.	29.93	29.90	29.92	30.44	30·44	30.38	29.26	29.36	28.98
Perrysburg, Ohio.	29.40	29.33	29.42	29.93	29.91	29.90	28.91	28.34	28.89
Pittsburg, Pa. (Oakland Station.)	28.95	28.93	28.94	29.53	29.49	29.43	27.96	27.94	28.15
Pittsburg, Pa. (Marine Hospital.) (1)	29.00	28.97	28.98	29.55	29.53	29.44	28.15	28.05	28.17
Pomfret, Conn.	28.90	28.89	28.90	29.35	29.35	29.34	28.12	28.23	27.95
Portland, Me	29.83	29.82	29.86	30.35	30.84	20.36	28.81	29.01	29.16
Pottsville, Pa.	29.29	29.22	29.26	29.81	29.77	29.70	28.57	28.22	28.40
Princeton, Mass.	28.80	28.80	28.80	29.30	29.31	29.29	27.91	28.10	27.76
Providence, R. I	29.82	29.79	29.83	30.33	30.31	30.32	28.94	29.10	29.38
Quasqueton, Iowa.	29.11	29.08	29.10	29.59	29.53	29.61	28.41	28.31	27.98
Sag Harbor, N. Y.	29.78	29.76	29.78	30.24	30.28	30.28	28.94	29.07	28.68
St. Louis, Mo.	29.54	29.51	29.52	29.97	29.98	29.99	28.87	28.89	28.60
St. Martins, C. E.	29.81	29.80	29.80	30.30	30.28	30.30	28.69	28.82	28.74
Savannah, Ga.	30.15	30.09	30.13	30.60	30.54	30.57	29.77	29.68	29.73
Smithsonian Inst.	30.14	30.08	30.10	30.74	30.68	30.60	29.48	29.07	29.18
Southwick, Mass. (2)	30.01	29.99	....	30.50	30.35	....	29.13	29.27	....
Sparta, Ga.	29.49	29.48	29.50	29.90	29.80	29.86	29.04	29.04	29.16
Spencertown, N. Y.	29.46	29.41	29.44	29.94	29.90	29.96	28.57	28.59	28.35
Springdale, Ky.	29.47	29.44	29.47	29.98	29.88	29.92	28.58	28.90	29.01
Steuben, Me.	30.02	30.03	30.02	30.72	30.74	30.70	29.19	29.34	29.41
Superior, Wis.	29.35	29.32	29.34	29.85	29.77	29.78	28.63	28.60	28.70
The Rock, Ga. (3)	....	....	....	....	....	....	....	....	....
Upper Alton, Ill.	29.48	29.48	29.49	29.97	29.79	29.87	28.74	28 98	28.66
Urbana, Ohio.	28.75	28.75	28.77	29.16	29.13	29.11	....	....	....
Warrington, Fa.	30.11	30.11	30·09	30.45	30.48	30.44	29.82	29.82	29.79
Westfield, Mass.	29.89	29.88	29.91	30.25	30.25	30.30	29.49	29.52	29.42
Williamstown, Mass.	29.26	29.25	29.26	29.76	29.80	29.76	28.35	28.47	28.20
Worcester, Mass.	29.48	29.47	29.47	30.24	29.93	29.93	28.65	28.74	28.95
Zanesville, Ohio.	29.18	29.17	29.19	29.74	29.61	29.64	28.24	28.38	28.58

(1) Obs. taken at Sunrise, 9 A. M. 3 P. M. and 9 P. M. Mean at 9 A. M. 29.01, Max. 29.55. Min. 28.06.

(2) Incomplete.

(3) Mean for the month 29·45 Max. 29.70 Min. 29.00, obs. taken morning 3 P. M. and evening,

NAME OF STATION.	Mean.			Maxima.					Minima.				
	7 A. M.	2 P. M.	9 P. M.	7 A.M	2 P.M	9 P.M	Warmest day. Date.	Warmest day. Mean temp.	7 A. M.	2 P. M.	9 P. M.	Coldest day. Date.	Coldest day. Mean temp.
Alexandria, Va.....	31.13	39.83	33.86	48	58	46	7	50.67	18	20	20	31	23.67
All Saints, S. C.....	43.55	56.00	47.55	61	71	69	21	66.67	26	42	31	27	35.33
Amherst, Mass......	24.76	31.63	26.77	51	48	42	29	41.33	11	13	11	14	11.90
Angelica, N. Y.......	20.84	30.90	24.32	50	43	46	6	40.00	0	16	6	23	10.50
Ann Arbor, Mich...	....	....	....	...	...	...	...	....	.....	.....	.....	...	....
(Woodruff).........	22.00	30.00	25.10	48	60	60	6	56.00	–1	10	–1	23	4.67
Ann Arbor, Mich....	....	....	....	...	...	...	...	....	.....	.....	.....	...	....
(Winchell).........	22.15	27.95	24.34	51	58	59	6	56.00	–1	8	–2	23	5.30
Arcola, Ohio.........	27.08	33.45	29.65	44	54	54	6	48.67	6	18	10	23	13 33
Ashland, Va.........	30.35	42.42	34.26	58	66	58	6	59.00	10	22	13	30	17.00
Athens, Ill...........	22.97	35.55	26.58	57	66	61	2	59.67	–11	11	3	22	3.33
Auburn, Ala.........	41.84	56.00	46.55	62	78	64	8	64.00	17	34	27	30	26.67
Augusta, Ill..........	22.00	33.09	25.80	59	65	60	2	60.00	–13	7	2	22	2.00
Austin, Texas........	40.74	60.13	47.26	64	81	62	2	64.33	16	32	28	30	26 67
Baldwinsville, N. Y.	25.19	29.26	25.25	47	45	39	7	42.00	2	10	8	14	6.67
Ballardsville, Ky....	30.64	38.26	33.52	58	64	62	6	61.33	8	16	13	30	13.00
Battle Creek, Mich.	22.39	31.58	23.55	50	60	50	7	52.67	1	10	0	23	6.33
Bedford, Pa..........	25 89	34.54	27.59	48	55	43	12	43.80	7	20	12	30	16.33
Beloit, Wis...........	18.41	27.93	21.58	47	57	54	2	51.67	–11	6	–4	23	–2.33
Beverly, N. Y........	27.52	33.06	29.45	50	47	44	7	42.00	13	16	15	14	14.67
Bladensburg, Md.(1)	31.03	40.29	35.17	48	60	56	17	41.50	18	20	22	31	23.83
Bloomfield, N. J....	29.37	35.72	30.60	53	52	43	7	44.16	14	17	15	14	16.17
Boston, Mass.........	26.68	36.42	29.68	56	56	49	7	45.67	13	20	10	14	14.33
Brandon, Vt..........	22.91	29.67	24.30	44	50	40	29	43.00	0	4	0	14	1.50
Burlington, Vt......	23 03	29.33	22.87	45	49	40	7	43.33	–6	6	–2	14	0.67
Burlington, N. J....	30.71	37.71	31.77	51	65	48	7	48.33	18	24	16	14	21.33
Camden, S. C........	37.64	54.90	43.71	56	70	63	21	60.33	18	40	25	31	29.67
Canton, N. Y........	20.13	27.35	22.55	52	48	46	7	48.00	–16	–5	1	14	–6.67
Carmel, Me.	16.89	26.52	21.52	36	49	46	29	40.00	–6	–2	–10	14	–3.33
Castleton, Vt.........	23.00	32.41	24.45	52	56	42	29	46.00	2	8	2	14	1.33
Cedar Keys, Fa.....	53.26	61.23	57.84	65	74	68	21	66.00	36	42	43	31	40.67
Ceresco, Wis.........	14.68	20.64	16.14	51	55	52	2	39.00	–14	–10	–11	13	–9.33
Chapel Hill, N. C....	35.26	49.52	40.58	50	70	60	7	58.67	19	33	22	30	28.00
Charleston, S. C.....	46.22	56.96	50.19	60	69	60	7	....	28	42	34	31	....
Cincinnati, Ohio.(2).	32.03	40.81	35.42	57	64	63	6	60.67	13	23	17	30	18.33
Columbus, Miss. (3)..	....	....	....	64	72	66	...	....	16	32	25	0	....
Concord, N. H......	20.13	29.13	23.94	44	48	42	22	40.00	2	12	2	14	8.00
Cooper, Mich.........	24.16	29.64	25.16	52	60	53	3	54.67	4	13	2	23	8.33
Crichton's Store, Va.	39.30	46.30	43.79	58	67	62	21	59.00	26	32	33	31	31.33
Danville, Ky.........	36.52	46.97	39.81	57	69	63	6	63.00	13	24	18	30	19.00
Deaf & Dumb Inst.	....	....	....	...	...	...	...	....	.....	.....	.....	...	....
N. Y................	29.45	34.40	30.71	51	46	44	7	42.10	15	18	17	14	16.90
Detroit, Mich........	25.81	31.29	27.84	52	58	58	6	56.00	4	17	5	23	8.67
Dubuque, Iowa......	21.29	27.96	23.40	54	60	58	2	57.00	–6	–2	–2	13	–2.67
Easton, Pa. (4).......	22.68	32.84	29.32	37	44	40	18	37 00	4	17	16	14	14.67
Exeter, N. H.........	20.87	30.19	24.26	42	47	46	29	40.67	4	6	2	1	14.33
Flint, Mich...........	20.70	30.30	23.10	46	58	54	6	52.67	4	13	1	...	....
Fort Madison,.......	....	....	....	...	...	...	...	....	.....	.....	.....	...	....
Iowa, (5)...........	21.32	32.65	26.87	56	66	61	2	61.00	–10	5	3	22	2.67
Frederick City, Md.	29.96	37.18	33.03	47	56	46	7	48.23	17	23	21	31	21.93
Fryeburg, Me. (6)...	....	....	....	...	...	...	...	....	–12	0	–9	14	–3.50
Gallipolis, Ohio......	30.71	43.58	35.50	55	64	58	6	58.67	7	22	13	30	17.00
Gardiner, Me.........	21.94	29.29	24.19	45	49	47	22	42.00	–10	3	0	14	4.00
Germantown, Ohio.	26.15	35.37	29.30	57	61	61	6	59.90	5	12	10	22	10.40
Gettysburg, Pa......	27.03	34.35	28.74	46	55	48	7	49.67	9	22	15	31	19.00
Glenwood, Tenn....	35.00	45.32	38.69	62	66	65	6	64.40	9	20	14	30	16.00

(1) Incomplete.
(2) Obs. taken at 6 A. M. 1 P M. and 9 P. M.
(3) Mean for the month 48.18
(4) Obs. taken at 6½ A. M. 12 M. 5 P. M. No obs. from the 2nd. to the seventh inclusive.
(5) Obs. taken at 6 A. M. 12 M. and 7 P. M.
(6) Incomplete, coldest day only given.

NAME OF STATION.	Mean.			Maxima.					Minima.				
	7 A. M.	2 P. M.	9 P. M.	7 A. M	2 P. M	9 P. M	Warmest day Date	Warmest day Mean temp.	7 A. M.	2 P. M.	9 P. M.	Coldest day. Date	Coldest day. Mean temp.
Gouverneur, N. Y.	21.90	28.06	24.03	47	52	40	7	46.33	-20	10	-16	14	-8.67
Grand Rapids, Mich.	22.81	31.58	25.84	52	61	58	2	53.33	8	10	5	13	12.33
Granville, Ohio......	28.48	35.48	32.06	56	62	57	6	57.33	7	15	13	22	13.67
Green Springs, Ala.	40.42	56.31	46.16	62	73	63	20	63.67	16	32	26	29	25.33
Great Falls, N. H...	21.13	32.25	24.06	44	46	46	29	41.00	3	6	0	14	3.33
Hamilton, C. W..(1)	28.90	....	29.86	51	...	53	6	50.00	9	...	15	24	12.00
Harrisburg, Pa......	29.44	34.38	31.59	39	43	41	4	41.00	19	26	22	14	23.33
Hillsborough, Ohio.	27.48	36.45	30.84	51	61	58	6	56.83	6	16	8	30	11.50
Jackson, Ohio, ...	28.61	41.16	31.45	55	65	58	6	58.67	3	17	8	30	13.33
Jacksonville, Fa...	50.32	62.67	52.94	65	74	65	21	66.33	33	40	36	31	37.00
Janesville, Wis......	15.82	26.38	18.58	48	56	54	2	50.33	-14	-6	-8	13	-6.00
Keo kuk, Iowa......	22.77	34.83	24.87	60	66	60	2	60.67	-6	8	2	22	4.67
Key West, Fa.......	64.10	69.64	....	71	76	...	21	73.00	52	59	...	23	56.50
Knox Hill, Fa. (2)...	42.11	58.83	46.71	65	73	61	25	62.67	26	39	34	29	33.33
Knoxville, Tenn.....	34.50	45.68	38.52	61	64	58	6	57.10	13	24	19	29	21.33
La Grange, Ga. (3)..	39.06	53.74	50.81	60	73	66	7	64.33	15	30	27	30	25.00
Lewisburg, Va......	33.03	43.10	36.52	54	60	52	7	50.67	16	23	19	30	22.33
Lima, Pa.	30.25	36.13	31.01	43	61	47	7	50.40	12	22	17	14	20.60
Lodi, N. Y...........	24.23	29.45	24.16	46	46	39	4	39.33	4	12	10	14	9.33
Londonderry, N. H.	20.87	30.84	24.61	47	48	47	7	39.33	5	17	1	14	9.00
Lowville, N. Y. (4)..	22.32	29.87	63.45	44	50	40	7	42.33	-8	-1	6	14	-1.00
Madison, Ind.........	30.60	38.21	33.32	56	61	61	6	59.50	10	19	12	30	14.16
Manchester, N. H...	21.00	31.54	27.48	51	51	46	22	44.33	4	15	6	14	9.00
Manchester, Ill......	24.64	34.35	28.10	56	68	60	2	59.00	-4	10	4	22	5.00
Marietta, Ohio......	28.94	39.52	32.45	58	62	56	6	56.33	11	15	14	23	15.67
Mendon, Mass.......	23.45	31.52	29.37	40	55	52	7	45.67	12	14	10	14	12.67
Millersburg, Ky.....	31.06	39.50	35.23	54	61	59	6	58.00	10	18	16.	30	15.00
Milton, Ind...........	27.77	35.03	32.05	54	59	59	6	57.50	0	12	5	22	8.16
Milwaukie, Wis.....	19.23	27.00	21.58	51	55	53	2	51.33	-10	-2	-8	23	-3.00
Monroe, Mich........	22.52	29.68	25.94	48	56	48	6	50.67	-4	12	4	23	6.67
Montcalm, Va. (5)...	37.18	45.60	42.77	46	58	55	7	53.00	30	39	38	2	36.67
Morrisville, Pa.	28.87	36.52	30.26	49	61	48	7	49.00	12	23	17	26	21 00
Moss Grove, Pa......	23.64	30.90	26.87	...	...	...	6	49.33	2	14	8	23	9.33
Mount Vernon, Ohio.	25.13	36.32	31.03	50	58	56	6	54.67	4	15	8	24	11.00
Muscatine, Iowa....	20.16	29.97	22.58	54	64	60	2	59.33	-23	2	-11	22	-4.67
Nantucket, Mass.....	35.85	39.48	36.61	50	53	48	7	49.50	23	23	18	14	31.33
Newark, N. J. (6)....	....	....	....	...	...	...	18	43.62	...	...	...	14	18.00
Newark, Ohio, (7)...	19.12	31.94	22.76	39	49	32	20	36.00	3	18	12	22	15.00
New Bedford, Mass.	28.53	36.50	31.08	47	54	48	7	49.00	13	17	10	14	14.80
Newburyport, Mass.	28.81	33.78	30.80	50	52	52	22	47.67	13	13	7	14	11.50
New Harmony, Ind..	30.61	40.29	34.63	58	65	57	2	57.33	9	20	11	22	16.33
New Lisbon, Ohio..	26.81	32.55	27.90	59	60	50	12	48.00	4	10	10	24	11.00
New Orleans, La....	50.34	60.29	54.39	65	73	67	6	68.33	29	37	37	30	34.83
New Wied, Texas...	44.90	62.10	48.94	67	83	63	20	65.33	20	39	30	30	30.67
Norristown, Pa......	29.50	35.53	30.62	41	46	41	13	40.10	14	25	20	25	22.00
North Attleboro,...	....	....	....	...	...	...	...	....	.....	.....	.....	...	....
Mass................	25.83	34.14	28.56	55	53	52	7	46.20	10	15	10	14	13.00
Norwalk, Ohio. (8)...	24.47	33.40	28.66	55	61	59	6	58.00	-2	13	3	23	7.00
Oberlin, Ohio........	25.70	33.52	29.03	52	58	57	6	55.67	4	20	8	23	11.00
Oldtown, Me. (9).....	20.43	27.05	24.05	40	47	42	22	40.00	3	2	4	14	3.00
Oswego, N. Y. (10)..	26.00	31.42	23.77	49	47	47	4	45.00	4	9	9	14	7.33
Ottawa, Ill............	21.81	29.32	24.16	52	62	56	2	55.67	-8	9	0	22	1.33
Oxford, Miss.........	38.42	49.55	43.26	62	67	63	6	62.70	11	26	18	30	17.00
Pella, Iowa...........	15.35	28.19	18.77	53	61	59	2	57.67	-8	0	-4	21	1.00
Penn Yan, N. Y(11).	25.48	31.74	26.81	49	47	40	4	42.67	6	19	10	14	12.00

(1) Obs. taken at 9 A. M. and 9 P. M.
(2) For the last 18 days of the month only.
(3) Obs. taken at Sunrise, Noon and Sunset.
(4) Obs. taken at 7 A. M. Noon and 9 P. M.
(5) For the first 11 days of the month only.
(6) Obs, taken at 7 A. M. and 5 P. M. Mean for the month 32.46, Max. 54.75, Min 12.
(7) For the last 18 days of the month only.
(8) Hours of observation not uniform.
(9) For the last 22 days of the month only.
(10) Obs. taken at 6 A. M. Noon and 6 P. M.
(11) Obs. taken at Sunrise, 2 P. M. and Sunset.

NAME OF STATION.	Mean. 7 A. M.	Mean. 2 P. M.	Mean. 9 P. M.	Maxima. 7 A.M	Maxima. 2 P.M	Maxima. 9 P.M	Warmest day. Date	Warmest day. Mean temp.	Minima. 7 A. M.	Minima. 2 P. M.	Minima. 9 P. M.	Coldest day. Date	Coldest day. Mean temp.
Perry, Me	24.35	29.19	27.16	44	45	46	7	42.67	–1	8	1	14	6.33
Perrysburg, Ohio (1)	26.48	35.39	29.68	56	69	65	6	63.33	0	19	6	23	8.33
Philadelphia, Pa.	32.00	37.10	33.90	45	62	50	7	52.33	17	22	21	14	21.33
Pittsburg, Pa. (Oak. Station)	27.61	35.77	30.23	56	56	57	16	50.00	8	18	13	23	14.33
Pittsburg, Pa. (Marine Hospital,) (2)	34.55	35.06	34.81	53	53	52	6	49.50	20	20	19	31	19.50
Platteville, Wis.	18.26	30.26	21.00	54	62	60	2	57.67	–10	–6	–12	13	–8.67
Pocopson, Pa.	29.50	37.00	30.90	44	61	43	7	49.00	9	23	21	25	20.67
Pomfret, Conn.	25.16	32.00	26.68	46	54	49	7	47.33	11	11	8	14	10.00
Pottsville, Pa.	25.77	33.00	28.19	43	55	42	7	46.67	11	17	18	14	15.33
Poultney, Iowa	16.32	27.80	21.45	52	58	58	2	55.67	–11	–4	–4	13	–6.33
Princeton, Mass.	23.17	27.67	23.39	48	48	46	7	44.50	7	8	5	14	6.67
Providence, R. I. (3)	26.50	35.00	28.40	46	54	47	7	44.33	12	16	10	14	12.67
Quasqueton, Iowa.	17.71	25.17	21.07	57	60	59	2	57.00	–8	5	1	22	–0.30
Randolph, Pa.	23.19	30.42	25.00	47	54	46	6	46.00	2	12	8	24	10.33
Richmond, Mass.	25.35	30.55	26.42	48	48	40	7	43.33	10	8	6	14	8.00
Sacramento, Cal. (Logan) (4)	....	....	....	...	...	...	...	....	...	...	...	...	....
Sacramento, Cal. (Hatch). (5)	....	....	....	...	...	...	25	53.25	...	...	...	6	34.26
Sag Harbor, N. Y.	31.06	39.55	32.81	53	59	50	7	52.00	17	18	13	14	16.00
St. James, Mich.	20.00	29.26	22.70	50	53	51	3	50.33	6	5	1	13	4.67
St. Johnsbury, Vt.	19.94	27.76	23.94	45	50	42	29	45.00	–4	–3	–1	14	–2.33
St. Joseph, Mich.	–3.52	3.39	2.39	34	32	21	16	20.33	–34	–15	–21	4	–19.00
St. Louis, Mo.	29.34	38.78	33.34	59	65	60	6	47.67	3	14	10	22	11.80
St. Martins, C. E.	15.10	23.19	17.71	37	38	36	7	36.67	–16	–6	–9	14	–11.00
Salmon Falls, N. H.	24.11	32.40	26.02	44	46	44	29	41.33	5	6	5	14	5.50
Savannah, Ga.	44.78	58.14	49.97	59	69	64	13	62.20	26	40	34	30	35.90
Saybrook, Conn.	29·68	34.39	30.32	49	48	46	7	45.33	15	16	15	14	15.33
Schellman Hall, Md.	28.48	38·68	30.78	50	58	49	7	52.33	13	25	15	27	20.00
Smithfield, Va. (6)	....	....	....	...	...	...	14	26.25	...	...	...	20	54.00
Smithsonian Inst.	31.85	39.65	35.73	47	57	57	7	51.00	20	26	22	25	26.00
Smithville, N. Y.	23.71	29.71	24.19	51	48	45	6	43.67	–8	10	8	14	3.33
Sparta, Ga.	39.23	54.61	45.16	60	74	61	7	63.67	20	33	26	22	27.67
Spencertown, N. Y.	24.87	30.77	25.55	51	46	38	7	41.33	5	8	12	14	9.33
Springdale, Ky.	30.42	41.16	37.39	59	65	64	6	62.83	10	18	16	22	16.17
Springfield, Mass.	25.00	33.10	26.69	52	50	38	29	42.33	8	12	8	14	9.33
Steuben, Me.	21.00	26.06	25.58	43	47	45	22	43.00	1	10	1	14	9.33
The Rock, Ga. (7)	....	....	....	...	...	...	...	....	...	...	...	...	....
Thornbury, N. C.	35.06	49.70	39.93	51	69	64	21	60.13	22	36	25	31	29.60
Tuscaloosa, Ala.	41.60	55.07	46.21	64	70	64	6	65.00	18	32	26	30	26.07
Upper Alton, Ill.	27.16	36.89	30.95	58	65	60	2	59.67	5	15	9	22	10.00
Wampsville, N. Y.	24.94	32.55	25.71	48	50	40	7	42.00	1	16	6	14	9.33
Warrington, Fa.	50.16	61.10	58.58	68	76	66	9	67.33	28	40	42	30	37.33
Westfield, Mass.	24.84	30.35	26.42	47	42	38	22	38.67	10	11	11	14	10.67
W. Haverford, Pa.	30.16	37.19	....	44	62	...	7	53.00	14	22	...	14	21.00
White Marsh Island	44.00	56.97	48.30	58	69	62	13	61.67	25	40	32	31	35.33
Williamstown, Mass	23.19	29.48	25.10	51	52	44	7	43.20	7	9	8	14	8.23
Winchester, Va.	29.25	41.32	32.03	46	58	48	4	48,67	14	28	16	30	21.67
Woodward H. S. O.	26.66	40.06	34.10	56	66	62	6	58.00	9	16	14	22	15.33
Worcester, Mass.	24.55	31.23	27.10	53	49	43	29	41.33	11	14	11	14	12.33
Zanesville, Ohio	27.06	37.42	30.42	50	64	57	6	50.67	8	17	10	30	15.00

(1) Obs. taken at 7 A. M. 1 P. M. and 9 P.M.
(2) Obs. taken at Sunrise. 3 P. M. and 9 P. M. Mean at 9 A. M. 32.54, Max. 54. Min. 14.
(3) Obs. taken at Sunrise, 1 P. M. and 10 P. M.
(4) Mean for the month 43.71, Max. 67. Min. 27.
(5) Mean for the month 44.74, obs. taken at Sunrise, Noon, Sunset and 10 P. M.
(6) Mean for the month 40.22, Max. 66, Min. 19.
(7) Mean for the month 45.55 Max. 65, Min. 25. Obs. taken morning, 3 P. M. and evening.

NAME OF STATION.	Mean.			Maxima.					Minima.				
							Warmest day					Coldest day.	
	7 A. M.	2 P. M.	9 P. M.	7 A.M.	2 P.M.	9 P.M.	Date	Mean temp.	7 A. M.	2 P. M.	9 P. M	Date	Mean. temp.
Alexandria, Va......	22.18	31.70	26.18	45	53	40	14	46.00	6	8	9	10	14.00
All Saints, S. C.....	38.71	51.57	43.10	62	71	61	8	62.67	26	36	27	27	30.33
Amherst, Mass......	15.53	24.49	19.61	38	42	35	15	38.33	–16	–7	–11	6	–8.00
Angelica, N. Y......	9.86	22.61	14.54	38	42	36	14	38.17	–28	–8	–19	6	–18.67
Ann Arbor, Mich...	....	....	....	...	...	...	...	....	.....	.....	.....	...	....
(Winchell).........	11.31	20.09	14.38	36	37	35	13	36.47	–14	0	–4	6	–6.07
(Woodruff).........	11.07	22.04	15.43	35	38	37	13	36.67	–24	3	–4	6	–5.67
Athens, Ill............	15.84	30.84	20.55	45	51	43	12	43.67	–6	10	1	25	2.33
Augusta, Ill..........	15.35	27.82	20.11	32	50	42	12	42.33	–10	7	–6	25	–1.67
Austin, Texas........	37·89	59.07	46.85	57	83	64	7	64.67	20	31	28	26	28.67
Auburn, Ala..........	35.64	51.64	43.32	62	70	62	8	60.67	20	30	27	26	25.67
Baldwinsville, N. Y.	12.68	20.93	16.79	32	34	33	14	33.67	–22	–11	–17	6	–16.67
Ballardsville, Ky...	23.00	31.64	27.75	46	54	48	12	47.00	2	17	12	26	10.67
Battle Creek, Mich.	12.61	28.11	16.04	41	43	38	13	40.33	–9	3	–4	6	–2.00
Bedford, Pa..........	16.92	26.25	19.96	35	55	37	14	38.67	–3	13	7	7	8.00
Beloit, Wis............	8.00	22.00	11.93	35	37	35	12	35.00	–14	5	–4	25	–3.33
Beverly, N. Y........	18.29	25.81	21.68	35	39	38	15	37.00	–12	–4	–9	6	–4.67
Bladensburg, Md.(1)	22.08	33.92	25.45	38	50	42	22	36.33	3	24	10	7	7.50
Bloomfield, N. J.....	19.48	28.79	22.80	38	43	38	16	37.17	–6	2	–9	6	–2.67
Boston, Mass..(2)....	19.28	29.89	22.57	39	52	40	15	40.67	–7	0	6	6	1.67
Brandon, Vt..........	12.98	23.47	16.15	34	41	35	15	36.33	–10	–14	–18	6	–17.83
Burlington, Vt......	11·28	22.93	15.61	33	44	38	15	38·33	–24	–13	–20	6	–19.00
Burlington, N. J....	20.61	30.71	25.64	48	48	46	14	47.00	–2	5	2	7	3.67
Camden, S. C........	31·46	52.30	39.71	56	69	59	7	54.33	16	36	25	27	28.00
Carmel, Me...........	7.46	19.76	10.68	33	41	34	17	35·33	–24	–15	–21	7	–12.67
Canton, N. Y.........	6.50	19.21	11.54	34	37	32	14	34.33	–40	–17	–28	6	–27.00
Castleton, Vt........	12.64	28.78	16.79	40	57	34	18	39.67	–22	–6	–17	6	–14.00
Cedar Keys, Fa......	48.39	55.25	52.75	60	64	62	8	62.00	32	32	44	28	40.00
Ceresco, Wis.........	1.40	17.23	7.51	26	34	24	12	24.67	–14	–10	–8	26	–8.67
Charleston, S. C.....	41.17	53.14	46.67	60	68	60	8	....	28	38	34	27	....
Chapel Hill, N. C...	28.79	45.54	34.86	51	65	54	14	53.00	18	31	23	28	25.00
Cincinnati, Ohio (3)..	25.68	33.04	30·68	46	52	47	12	46.00	7	21	14	26	14.00
Columbus, Miss (4)..	....	....	....	54	67	63	...	....	15	31	25	...	....
Concord, N. H......	12·35	25.21	17.21	34	40	34	15	35·33	–24	–4	–16	6	–11.33
Cooper, Mich.........	17.86	23.03	19.28	40	40	40	13	38.67	4	8	0	7	7.03
Deaf & Dumb Inst...	....	....	....	...	...	...	...	....	.....	.....	.....	...	....
N. Y..................	20.87	27.27	23.35	37	41	39	15	37.41	–7	3	–8	6	–4.73
Danville, Ky.........	30.00	41.21	35.32	52	69	56	7	58.67	11	24	21	26	18.67
Detroit, Mich........	13.82	23.39	16.82	38	40	36	13	38.00	–14	–3	–4	6	–7.00
Dubuque, Iowa......	13.43	24.15	18.67	37	39	33	12	35.33	–8	7	–1	25	–0.67
Easton, Pa............	18.71	29.29	20.60	37	48	36	14	37.67	–13	0	–6	6	–5.00
Exeter, N. H.........	12.32	23.11	15.94	36	38	35	15	35.33	–20	–8	–14	6	–11.33
Flint, Mich...........	8·40	22.70	13.80	35	38	36	14	36·33	–31	4	–9	...	....
Fort Madison, Iowa.	14·57	27.64	21·25	40	48	44	12	42.67	–11	8	–2	25	–0.67
Frederick, Md......	22.52	30.62	26.03	41	46	43	14	41.10	4	7	5	7	5.43
Fryeburg, Me.. (1)...	–8·64	23·13	·11·25	...	...	...	...	....	–38	–6	–23	7	–15.67
Gallipolis, Ohio.....	22.89	37.89	26·68	40	57	48	13	43.33	7	23	14	26	14.67
Gardiner, Me.........	14.57	23.89	18.45	38	41	35	19	36.33	–18	–9	–12	6	–11.00
Garlandsville, Miss.	....	....	....	...	...	...	...	....	.....	.....	.....	...	....
...(5)	40.43	56.14	44.46	62	75	76	7	65.00	26	34	29	25	31.33
Germantown, Ohio.	17.82	27.95	21.87	39	46	44	12	41.67	-0.6	13	5	26	5.9
Gettysburg, Pa......	17.71	28.64	21.57	37	46	37	14	38.33	–5	3	0	7	1.33
Glenwood, Tenn.....	28.14	40.50	34.42	58	59	59	7	58.50	8	21	16	26	15.77
Gouverneur, N. Y...	6.43	19.54	15.84	35	40	38	14	35·67	–44	–20	–24	5	–26.00
Grand Rapids, Mich.	12.00	25.54	16.36	39	40	36	13	37.67	–8	9	–2	6	–0.67
Granville, Ohio......	18.39	26.68	23.39	40	47	44	13	42.33	3	10	10	26	7.26

(1) Incomplete.
(2) Obs. taken at 6 A. M. 2 P. M. and 10 P. M.
(3) Obs. taken at 6 A. M. 1 P. M. and 9 P. M.
(4) Mean for the Month 42.64,
(5) Obs. taken at 7 A. M. 1 P. M. and 9 P. M.

NAME OF STATION.	Mean.			Maxima.					Minima.				
	7 A. M.	2 P. M.	9 P. M.	7 A.M.	2 P.M.	9 P.M.	Warmest day Date	Warmest day Mean temp.	7 A. M.	2 P. M.	9 P. M.	Coldest day. Date	Coldest day. Mean temp.
Great Falls, N. H...	12.93	27.46	15.83	37	42	33	16	37.33	-24	-4	-17	6	-11.67
Green Springs, Ala.	35.25	52.83	42.22	62	68	64	7	61.67	20	35	26	26	27.00
Hamilton, Ca..(1)...	18.86	....	19.43	38	...	38	14	38.00	-15	.....	-5	6	-10.00
Harrisburg, Pa......	22.21	29.14	25.88	41	43	40	14	39.33	-1	0	4	7	1.33
Hillsborough, Ohio.	18.38	28.18	24.18	38	49	43	12	41.00	2	15	9	26	8.83
Janesville, Wis.....	7.29	20.39	10·36	36	40	40	12	35.33	-19	4	-8	3	- 3.00
Jackson, Ohio.......	20.18	31.89	24.61	39	53	44	13	44.33	5	18	14	26	13.67
Jacksonville, Fa....	45.71	60.03	49.57	64	73	66	8	69.00	30	48	38	27	40.67
Key West, Fa........	62.14	68.75	....	72	78	...	19	75.00	54	59	.....	1	59.00
Knoxville, Tenn.....	30.51	42.83	36.05	53	62	56	7	54.67	12	23	18	26	18.33
Knox Hill, Fa.	41.43	56.82	46.85	65	72	68	23	61.33	27	38	32	26	32.33
Lac qui parle, Min.	5.14	14.93	9.14	28	38	26	19	28.00	-19	-10	-14	23	-13.33
Lewisburg, Va......	25.17	36.14	28.93	38	53	44	13	40.67	12	22	17	27	17.00
Lima, Pa..............	20.80	29.17	24.58	43	45	38	14	42.03	-1	2	2	6	3.63
Lodi, N. Y...........	12.11	20.93	14.57	35	41	35	14	37.00	-22	-11	-22	6	-18.33
Londonderry, N H.	13.03	23.14	16.89	35	40	35	16	36.00	-23	-10	-17	6	-14.00
Lowville, N. Y ..(2)..	8.96	22.03	14 71	35	43	36	15	35.67	-38	-12	-28	6	-26.00
Madison, Ohio.......	16.89	24.60	16.78	38	37	38	13	36.33	- 8	4	2	6	0.00
Madison, Ind.........	25.89	33.11	29.37	45	51	48	12	46.00	7	20	14	26	13.67
Manchester, N. H..	11.85	27.32	20.59	36	42	37	16	37.33	-22	-12	-18	6	-15.33
Manchester, Ill (3)...	18.43	28.37	23.29	44	46	44	12	44.00	-3	9	1	25	3.67
Marietta, Ohio........	19.00	29.90	24.43	39	46	44	13	42.67	6	18	13	26	12.33
Mendon, Mass......	15.64	23.29	18.50	34	39	37	15	36.00	-17	-11	-15	6	-10.33
Millersburg, Ky.....	24.00	34.18	28.89	44	58	48	12	46.67	5	18	15	26	12.67
Milton, Ind..........	18.52	29.00	23.07	39	45	42	12	40.50	-1	14	7	26	7.33
Milwaukie, Wis......	8.61	21.32	13.57	34	36	34	13	33.00	-10	4	-2	25	- 2.33
Monroe, Mich........	13.25	22.21	18.57	32	38	33	13	34.33	-10	4	0	6	- 1.33
Mount Vernon, O...	15.53	27.57	20.78	38	44	42	13	40.67	0	11	7	6	7.33
Morrisville, Pa......	20.21	30.94	21.71	40	46	38	14	40.00	-4	2	0	6	3.33
Muscatine, Iowa....	11.61	32.86	15.07	40	40	32	12	36.33	-11	3	5	25	- 3.00
Nantucket, Mass....	27.93	31.96	27.71	44	44	40	15	42.50	11	8	4	6	8.33
Newark, Ohio ..(4)...	18.36	30.48	25.00	36	48	44	13	41.33	4	16	15	3	12.25
Newark, N. J...(5)...	....	....	....	...	...	46	16	37.87	-8	.....	.....	7	2.25
New Bedford, Mass.	19.36	28.82	22.43	39	42	38	15	39.17	-12	-6	-14	6	-7.83
Newburyport, Mass.	17.54	25.63	19.55	42	41	38	15	38.70	-13	-4	-8	6	-6.00
New Harmony, Ind..	23.89	37.00	28.64	55	58	48	12	50.33	6	20	13	26	13.33
New Lisbon, Ohio...	14.68	22.54	17.82	35	39	32	14	33.67	5	7	5	26	8.67
New Wied, Texas...	42.82	60.46	48.50	63	87	65	5	68.00	22	32	28	26	29.67
New Orleans, La...	44.20	54.68	48.89	59	70	64	7	63.33	29	36	35	26	34.67
Norristown, Pa.....	22.13	28.18	24.89	41	44	38	14	40.67	-2	-3	2	6	3.33
North Attleboro,....	....	....	....	...	...	...	...	....	.....	.....	.....	...	....
Mass................	16.30	26.38	19.83	42	41	38	15	40.43	-15	-9	-12	6	-9.00
Oberlin, Ohio........	15.26	27.85	18.00	40	44	42	13	42.00	-10	10	2	6	2.33
Oldtown, Me.........	9.82	20.00	13.75	31	42	32	17	35.00	-18	-12	-18	6	-16.00
Oswego, N. Y.......	12.96	22.61	17.53	34	36	32	15	33.33	-22	-6	-21	6	-14.00
Ottawa, Ill...........	13.17	26.89	16.03	38	44	39	12	39.33	-7	10	0	25	2.67
Oxford, Miss........	31.77	46.02	39.18	54	66	62	5	57.83	10	22	16	27	18.67
Pella, Iowa..........	10.53	26.18	10.46	34	48	29	10	31.33	-10	5	- 7	25	-3.67
Penn Yan, N. Y..(6)	12.93	25.71	18.43	36	45	37	14	39.33	-19	-7	-16	6	-14.00
Perry, Me..........	14.32	22.18	17.14	35	38	34	17	34.67	-15	-4	-10	6	-8.00
Perrysburg, Ohio (3)	16.82	30.61	20.96	43	48	44	13	44.33	-7	7	0	6	3.67
Philadelphia, Pa....	23.10	30.00	26.90	43	46	41	14	43.33	-1	3	3	6	4.50
Pittsburg, Penn....	....	....	....	...	...	...	...	....	.....	.....	.....	...	....
(Oakland Station.)	15.96	24.68	19.89	37	52	40	15	36.67	-9	11	8	6	3.33
(Marine Hospital.)	23.48	24.89	25.14	38	39	40	13	37.67	10	11	15	27	13.00
Platteville, Wis.....	8.46	24.61	13.61	34	40	34	12	34.67	-14	6	-10	25	-6.00

(1) Obs. taken at 9 A. M. and 9 P. M.
(2) Obs. taken at 7 A. M. Noon and 9 P. M.
(3) Obs. taken at 7 A. M. 1 P. M. and 9 P. M.
(4) Incomplete. (5) Obs. taken at 7 A. M. and 5 P. M. Mean for the month 25.92.
(6) Obs. taken at Sunrise, 2 P. M. and Sunset.

NAME OF STATION.	Mean.			Maxima.					Minima.				
							Warmest day					Coldest day.	
	7 M. A.	2 P. M.	9 P. M.	7 A.M.	2 P.M.	9 P.M.	Date	Mean temp.	7 A. M.	2 P. M.	9 P. M.	Date	Mean temp.
Pocopson, Pa........	20.38	30.04	24.83	43	44	38	14	41.33	–2	2	2	7	4.33
Pomfret, Conn......	16.07	24.41	18.46	37	39	35	15	37.00	–16	–9	–14	6	–9·00
Pottsville, Pa.......	19.88	26.33	21.14	36	41	37	14	37.33	–6	–3	–6	6	–4.00
Poultney, Iowa......	10.15	21.11	11.85	33	42	28	12	32·00	–12	–1	–9	25	–7.33
Princeton, Mass.....	14.18	20.26	16.93	35	36	33	16	34.33	–17	–4	–2	7	–5.57
Providence, R. I (1)	18.60	27.30	20.50	42	42	36	15	40.00	–15	–7	–14	6	–8.00
Quasqueton, Iowa..	9.14	23.52	13.14	33	42	30	12	32.17	–11	4	–5	25	–4.27
Randolph, Pa.......	10.75	20.68	13.25	38	38	31	14	35.00	–22	4	–5	6	–11.33
Sacramento,.........	....	....	....	...	...	...	...	....	.....	.....	.....	...	....
(Logan)Cal.(2)....	....	....	....	...	...	...	...	....	.....	.....	.....	...	....
Sacramento,.........	....	....	....	...	...	...	...	....	.....	.....	.....	...	....
(Hatch)Cal·(3)....	....	....	....	...	...	...	15	58.00	.....	.....	.....	20	45.25
Sag Harbor, L. I...	21.32	30.71	23.93	35	46	40	15	40.33	–4	2	–6	6	–0.33
St. James, Mich....	7.71	21.55	11.07	30	41	31	21	31.00	-14	–6	–16	5	–9.83
St. Joseph, Min.....	–0.79	15.21	2.68	25	46	24	15	30.67	–20	–10	–20	22	–16.67
St. Louis, Mo.......	24.77	35.20	29.64	55	65	49	6	47.00	5	15	9	25	7.33
St. Martins, Ca. E..	3.68	17.53	9.23	32	38	34	16	34.53	–34	–19	–24	6	–25.23
Savannah, Ga.......	40.10	54.70	46.90	63	74	64	8	66.97	27	34	34	27	32.73
Smithfield, Va (4)...	....	....	....	...	...	...	22	42.25	.....	.....	.....	10	16.50
Schellman Hall, Md.	19.46	32.00	23.43	43	50	41	14	43.33	–2	4	4	7	2.33
Smithsonian Inst (5)	24.66	32.29	23.92	45	53	37	22	38.50	4	6	8	7	6.50
Smithville, N. Y....	7.53	20.25	13.35	38	42	35	14	38.00	–38	–12	–14	6	–26.67
Sparta, Ga..........	34.38	51.46	40.61	61	68	59	8	60.67	18	33	26	27	26.00
Spencertown, N. Y..	14.14	21.32	17.25	38	43	44	15	41.33	–19	–12	–14	6	–13.33
Springdale, Ky......	22.59	35.53	31.35	44	57	49	12	48.17	3	22	19	26	14.00
Springfield, Mass...	15.03	27.57	19.00	36	42	36	15	36.33	–19	0	12	6	–6.00
Steuben, Me.........	13.29	20.79	16.29	35	37	34	17	35.33	–19	–10	–15	6	–9.33
Taunton, Mass......	19.96	26.14	22.25	44	46	42	16	44.00	–12	– 6	–10	6	–6.67
The Rock, Ga. (6)...	....	....	....	...	...	...	...	....	.....	.....	.....	...	....
Thornbury, N. C...	28.26	43.63	33.23	51	66	53	14	54.67	16	28	20	27	24.20
Tuscaloosa, Ala.....	34.69	50.75	40.20	60	66	66	7	60.33	19	31	24	27	25.93
Upper Alton, Ill....	21.39	34.18	27.84	48	53	50	12	47.67	2	15	6	25	9.33
Urbana, Ohio........	14.61	30.93	20.18	40	52	42	13	41.67	–5	10	3	26	3.33
Wampsville, N. Y...	12.34	24.59	15.93	34	41	35	14	36.00	–27	–12	–24	6	–21.00
Warrington, Fa.(7)..	44.29	56.11	52.68	68	72	64	23	64.67	27	40	40	26	35.67
Westfield, Mass	15.29	23.04	18.79	35	38	35	15	35.67	–15	–7	–11	6	–7.33
W.Haverford, Pa.(8)	21.32	30.50	....	43	46	...	14	44.50	–4	0	.....	7	–2.00
White Marsh Is, Ga.	40.07	53.68	46.04	62	75	62	8	66.33	28	37	32	27	32.33
Williamstown, Mass.	....	....	....	...	...	...	...	....	.....	.....	.....	...	....
...(9)	14.52	20.77	16.50	37	40	34	15	37.17	–18	–13	–15	6	–14.17
Woods Hole, Mass..	22.68	28.71	25.57	39	43	39	15	40.33	–2	–4	–7	6	–2.33
Winchester, Va......	21.82	33.79	24.68	40	52	40	14	42.67	7	12	6	7	8.33
Woodward H.S.O.(10)	22.68	32.28	28.68	43	52	48	12	45.67	4	18	13	26	12.33
Worcester, Mass.....	15.36	23.57	18.68	36	39	37	16	37.00	–16	–10	–14	6	–10.33
Zanesville, Ohio.....	18.36	26.89	23.04	38	48	45	14	42.00	4	13	10	26	9.33

(1) Obs. taken at Sunrise, 1 P. M. and 10 P. M.
(2) Mean for the month 52.5, Max. 70, Min. 32.
(3) Obs. taken at Sunrise, Noon, Sunset and 10 P. M. Mean for the month 52.82.
(4) Mean for the month 30.73, Max. 65, Min. 10.
(5) Obs. taken at 9 A. M. 2 P. M. and 9 P. M.
(6) Mean for the month 40.57, Max 63, Min 24, obs. taken morning, 3 P M and evening.
(7) Obs. taken at Sunrise, Noon and Sunset.
(8) Incomplete.
(9) Obs. taken at 7½ A. M., 2 P. M. and 9 P M.
(10) Obs. taken at 7 A. M., Noon and 9 P. M.

NAME OF STATION.	Mean.			Maxima.					Minima.				
	7 M. A.	2 P. M.	9 P. M.	7 A.M	2 P.M.	9 P.M.	Warmest day Date.	Warmest day Mean temp.	7 A. M	2 P. M	9 P. M	Coldest day. Date.	Coldest day. Mean temp.
Alexandria, Va.....	33.6	45.5	38.4	52	64	55	6	55.3	17	34	24	1	25.0
All Saints, S. C.....	46.1	60.1	51.8	68	81	70	17	72.7	25	42	31	1	33.0
Amherst,.Mass......	27.4	37.2	30.0	42	57	40	31	44.2	7	25	16	1	19.0
Angelica, N. Y......	23.0	36.4	26.0	40	51	43	5	43.0	–8	18	14	1	9.0
Ann Arbor, Mich...	....	....	....	...	...	...	...	....	...	...	...	...	....
(Winchell,).......	24.4	33.6	28.9	39	51	40	5	42.3	5	20	14	24	16.0
Ann Arbor, Mich...	....	....	....	...	...	...	...	....	...	...	...	...	....
(Woodruff)........	24.1	35.5	29.8	39	54	41	31	44.0	3	23	16	1	16.3
Ashland, Va.........	33.6	47.1	36.9	58	66	58	17	54.0	8	26	18	1	22.0
Athens, Ill...........	28.6	43.6	33.5	52	68	52	5	56.0	2	27	12	20	15.7
Auburn, Ala.........	46.2	63.6	54.6	70	82	72	16	73.7	23	44	37	1	34.7
Augusta, Ill..........	27.2	40.9	33.2	51	65	52	8	53.0	–1	18	13	20	12.3
Austin, Texas.........	45.4	70.5	54.5	68	98	76	8	77.0	24	46	38	21	40.0
Baldwin Inst. Ohio.	25.6	34.9	30.1	44	51	45	31	44.3	–8	14	15	1	10.7
Baldwinsville, N. Y.	25.2	33.7	28.4	40	48	43	5	43.0	10	16	14	10	13.3
Ballardsville, Ky...	34.0	42.8	38.4	54	54	58	11	52.7	14	29	24	28	24.7
Battle Creek, Mich.	26.1	39.9	28.2	43	58	40	5	47.0	–5	26	10	1	12.3
Bedford, Pa..........	29.1	41.3	33.6	44	59	54	6	46.3	8	27	16	1	18.5
Beloit, Wis...........	21.8	35.8	26.9	35	54	38	8	41.7	–1	20	3	20	8.3
Beverly, N. Y........	29.6	38.8	33.4	44	59	46	31	47.3	13	26	18	1	20.0
Bladensburg, Md.(1)	34.0	48.6	38.1	57	69	54	...	....	14	36	25	...	....
Bloomfield, N. J....	30.5	41.6	34.3	47	58	46	6	46.5	15	30	19	1	24.2
Boston, Mass..(2)...	28.3	41.4	32.4	45	54	50	31	50.3	13	32	25	1	24.7
Brandon, Vt.........	23.3	33.5	26.6	38	63	41	31	46.7	0	15	15	10	14.0
Burlington, N. J....	33.5	43.6	38.4	44	62	52	31	51.3	18	30	20	1	23.7
Burlington, Vt......	22.9	35.0	27.3	38	51	42	31	43.7	–4	16	14	10	12.7
Camden, S. C.	43.6	62.0	51.2	67	85	73	17	72.6	17	40	28	1	31.3
Canton, N. Y........	20.9	32.1	25.8	40	50	50	31	46.0	0	14	6	10	6.7
Carmel, Me..........	21.2	34.9	22.8	34	55	40	31	39.7	0	16	9	14	14.7
Castleton, Vt.........	24.9	38.9	28.5	38	59	41	31	45.7	0	18	12	10	13.3
Cedar Keys, Fa.....	56.1	63.6	60.0	70	74	70	18	71.3	35	49	43	1	42.7
Ceresco, Wis.........	20.5	34.4	21.6	34	48	34	5	37.3	0	18	–2	21	6.0
Charleston, S. C. ...	48.7	61.7	54.3	66	77	68	17	....	28	44	40	1	....
Chapel Hill, N. C...	39.2	55.7	45.6	63	75	64	6	63.6	19	39	26	1	30.0
Cincinnati, Ohio..(3)	35.9	46.2	39.0	50	63	50	13	50.7	18	32	24	28	28.0
Concord, N. H......	24.2	37.2	29.0	38	53	42	31	42.3	0	26	18	1	16.0
Columbus, Miss..(4).	....	....	....	64	82	76	5	....	21	44	31	...	....
Cooper, Mich.........	28.1	34.9	30.4	42	52	42	31	42.7	14	18	16	24	16.7
Crichton's Store, ...	....	....	....	...	...	...	...	....	...	...	...	...	....
Va.(5)................	42.1	51.6	51.8	60	66	68	6	63.7	28	35	37	22	33.0
Danville, Ky.........	38.8	51.2	42.7	59	70	61	11	63.3	17	35	28	1	28.0
Deaf & Dumb Inst.	....	....	....	...	...	...	...	....	...	...	...	...	....
N. Y.................	31.9	40.0	34 8	45	56	47	30	47.9	17	28	21	1	23.8
Detroit, Mich........	27.1	35.9	31.1	40	54	44	31	46.0	4	20	20	1	17.7
Dubuque, Iowa......	26.8	37.0	30.2	43	60	45	8	45.0	5	19	10	20	11.3
Easton, Pa...........	30.23	43.68	33.00	52	68	58	31	53.0	10	32	26	1	24.33
Exeter, N. H.........	25.7	37.2	27.9	42	54	42	31	42.0	7	22	19	1	18.0
Flint, Mich...........	24.10	39.50	30.40	43	55	42	5	46.67	–1	24	17	...	....
Ft. Madison, Iowa(6)	26.6	40.5	34.5	48	62	52	8	51·6	5	16	13	20	11·3
Frederick, Md.......	33.2	44.2	37.7	45	61	49	6	51.3	15	33	24	1	25.6
Gallipolis, Ohio......	35.6	47.4	37.9	58	65	55	17	53.3	16	27	21	1	24.7
Gardiner, Me........	25.4	36.8	28.6	38	52	42	31	43.67	8	22	21	1	20.0
Garlandsville, Miss.	47.8	69.1	55.4	72	87	76	16	78.0	25	50	30	1	38.3
Germantown, O.....	29.5	40.1	32.6	49	56	50	14	39.0	7	25	18	20	33.6
Gettysburg, Pa......	30.0	42.2	33.9	42	64	49	31	48.0	10	32	19	1	21.3
Glenwood, Tenn.....	37.1	51.4	43.5	65	74	65	12	67.2	17	32	26	1	27.0

(1) Obs. too irregular to show the warmest or the coldest day.
(2) Obs. taken at 6 A. M. 2 P. M. and 10 P. M.
(3) Obs. taken at 6 A. M. 1 P. M. and 9 P. M.
(4) Mean for the Month 53.89.
(5) Obs. taken at 9 A. M., Noon and 5 P. M., irregular.
(6) Obs. taken at 6 A. M., Noon and 7 P. M.

NAME OF STATION.	Mean.			Maxima.					Minima.				
	7 A. M.	2 P. M.	9 P. M.	7 A.M.	2 P.M	9 P.M.	Warmest day Date	Warmest day Mean temp.	7 A. M	2 P. M	9 P. M	Coldest day. Date.	Coldest day. Mean temp.
Gouverneur, N. Y...	23.5	31.8	27.1	40	50	38	5	42.0	0	8	10	1	6.0
Grand Rapids, Mich.	23.8	34.9	28.7	34	48	40	5	41.7	0	20	14	20	19.0
Granville, Ohio......	30.9	40.7	35.2	46	59	49	5	48.0	9	25	20	20	21.7
Green Springs, Ala.	47.1	66.6	55.2	72	87	75	16	77.3	23	46	30	1	33.7
Great Falls, N. H...	24.5	38.0	27.9	36	55	36	31	42.5	5	23	18	1	19.0
Hamilton, Ca...(1)...	31.2	....	33.0	46	...	47	31	46.5	13	0	21	24	17.0
Harrisburg, Pa......	33.2	42.6	37.6	44	60	51	31	50.3	18	34	25	1	27.7
Hillsborough, Ohio.	30.8	40.8	34.7	46	57	51	13	47·3	11	24	20	20	22.2
Key West, Fa.......	66.7	72.3	....	74	82	...	17	78.0	48	54	...	2	52.0
Jackson, Ohio.......	....	......	......	...	...	...	...	......	...	...	...	...	....
(Crookham.)......	34.1	46.6	36.5	57	64	56	13	51.3	9	29	21	1	24.3
Jacksonville, Fa.....	55.2	69.3	57.0	75	86	78	15	76.3	33	49	36	1	39.3
Knox Hill, Fa.......	51.7	65·9	56.0	73	82	73	15	76.0	28	43	36	1	37.0
Knoxville, Tenn.....	37.6	53.4	45.0	63	73	66	12	67.5	18	31	24	28	26.7
Lacquiparle, Minn..	20.1	32.7	24.1	40	52	39	4	40.2	–6	8	5	19	5.0
Lebanon, Tenn......	39.8	52.0	46.7	64	72	65	12	65.3	21	32	28	1	30.5
Lewisburg, Va.......	34.8	49.3	40.7	52	66	56	12	55.0	16	28	22	28	25.3
Lima, Pa.............	31.9	42.5	35.3	44	61	49	6	49.1	16	20	20	1	18.9
Lodi, N. Y............	25.6	34.9	27.0	44	57	43	31	47.3	10	11	12	10	13.0
Londonderry, N. H.	25.2	37.3	29.2	40	58	43	31	46.	9	26	15	1	21.0
Lowville, N. Y.(2)...	21.8	34.3	26.5	36	56	40	31	43.7	5	13	8	10	8.7
Madison, Ind.........	34.2	42.9	38.1	53	51	54	11	51.0	18	31	24	1	28.0
Manchester, N. H...	24.6	40.5	31.9	42	59	48	31	47.0	4	30	24	1	20.0
Manchester, Ill..(3)..	30.1	40.9	34.6	50	66	54	5	54.7	8	21	14	20	14.7
Marietta, Ohio.......	29.9	43.6	36.0	55	58	54	15	48.3	4	24	18	1	20.0
Mendon, Mass........	26.6	35.0	30·1	43	46	42	5	40.7	11	25	21	1	20.7
Millersburg, Ky.....	33.7	45.0	38.0	54	60	54	13	54.0	11	28	24	1	22.3
Milton, Ind...........	28.8	38.9	33.1	45	54	47	5	47.2	8	23	18	20	19.7
Milwaukie, Wis.....	23.3	33.7	26.9	37	47	40	5	40.3	0	18	7	21	9.0
Morrisville, Pa.	30.5	43.5	34.5	42	64	46	31	50.0	14	33	22	1	23.3
Moss Grove, Pa.....	24.6	35.3	28.7	40	52	48	14	44.3	0	18	12	...	....
Mount Vernon, O...	28.5	40.7	32.5	51	53	50	5	46.7	3	25	19	1	19.7
Muscatine, Iowa....	24.4	38.0	28.5	40	63	46	8	46.0	2	14	0	20	5.7
Nantucket, Mass...	35.0	39.0	35.2	44	52	44	31	45.8	28	32	26	1	29.8
Newark, Ohio....(4)..	30.2	42.8	34.6	50	60	48	13	46.5	7	25	20	1	19.0
Newark, N. J. ..(5)..	....	....	....	...	...	...	6	52.4	...	...	...	1	27.0
New Bedford, Mass.	29.7	40.4	31.3	44	58	41	31	44.8	12	27	19	1	23.0
Newburyport, Mass.	29.6	38.4	31.4	43	61	43	31	48.6	14	24	22	1	23 5
New Harmony, Ind.	33.5	45.8	37.3	56	63	58	11	59.0	11	30	23	1	23.3
New Lisbon, Ohio...	27.6	36.8	31.6	40	55	49	14	42.3	1	19	10	1	12.3
New London, Ct.....	31.7	42.5	32.9	44	57	45	31	46.7	16	32	19	1	24.7
New Orleans, La...	52.1	64.8	58.0	68	82	73	16	75.0	31	41	37	1	36.3
New Wied, Texas...	51.6	72.6	55.2	71	99	81	8	80.3	30	45	34	21	39.7
Norristown, Pa......	32 9	42.7	35.9	45	62	53	31	50.0	15	27	20	1	20.8
North Attleboro,....	....	......	......	...	...	...	...	......	...	...	...	...	....
Mass.................	28.2	39.0	30.4	42	60	42	31	45.1	11	27	17	1	22.6
Oberlin, Ohio....(6)..	28.4	39.1	32.5	43	64	46	5	51.0	–2	24	18	1	8.0
Oldtown, Me.........	18.4	31.8	25.8	34	52	38	31	38.3	8	20	11	22	16.3
Orange Hill, Fa.....	....	....	....	...	...	...	...	....	...	...	...	...	....
Oswego, N. Y.......	26.2	35.7	28.6	47	56	42	5	44.0	10	15	11	10	13.7
Ottawa, Ill...........	27.0	40.2	30.9	41	59	43	8	45.0	–1	20	10	20	12.7
Oxford, Miss.........	42.7	58.0	48.6	65	78	71	11	69.7	23	37	30	1	30.3
Pella, Iowa..........	22.8	38.5	27.0	40	60	44	8	46.3	1	18	7	20	8.7
Penn Yan, N. Y.(7)	27.26	35.58	29.45	44	53	42	5	44.33	14	16	16	10	17.33
Perry, Me...........	22.7	34.9	26.4	38	46	36	31	37.3	14	18	14	14	15.7
Perrysburg, Ohio(3)	31.6	41.4	32.9	45	59	56	5	48.3	2	22	18	1	19.0

(1) Obs. taken at 9 A. M. and 9 P. M.
(2) Obs. taken at Sunrise, Noon and 9 P. M.
(3) Obs. taken at 7 A. M. 1 P. M. and 9 P. M.
(4) No obs. on 1st at 9 P. M.
(5) Mean for the month 36.734, Max. 61.75, Min. 14.25.
(6) No obs. on 1st at Noon.
(7) Obs. taken at Sunrise, 2 P. M. and Sunset.

NAME OF STATION.	Mean.			Maxima.					Minima.				
	7 A. M.	2 P. M.	9 P. M.	7 A.M	2 P.M	9 P.M	Warmest day Date.	Warmest day Mean temp.	7 A.M	2 P.M	9 P.M	Coldest day. Date	Coldest day. Mean temp.
Pittsburg, Pa..........	....	......	......	...	...	...	...	......	...	...	...	...	....
(Oakland Station.).	28.1	38.7	33.3	46	56	54	5	48.7	5	20	14	1	19.3
Pittsburg, Pa.	....	......	......	...	...	...	...	......	...	...	...	...	....
(Marine Hospital.)(1	35.6	40.9	37.6	48	61	54	17	50.5	15	29	23	1	23.0
Philadelphia, Pa......	33.6	43.6	39.2	47	64	51	31	52.8	16	30	29	1	25.0
Platteville, Wis.........	23.9	36.2	28.2	48	50	42	31	43.0	–9	20	5	21	6.7
Pocopsion, Pa..........	32.1	42.7	35.7	48	61	52	6	51.2	15	32	20	1	22.7
Pomfret, Conn.........	27.0	36.8	29.2	44	57	41	31	44.3	11	27	13	1	19.7
Pottsville, Pa	28.6	39.6	29.8	42	65	38	31	43.0	9	28	16	1	17.9
Poultney, Iowa.........	22.0	34.5	27.2	40	48	40	8	41.3	2	11	4	20	5.7
Princeton, Mass.......	25.0	32.5	27.0	39	54	40	31	44.2	10	23	11	1	17.4
Providence, R. I.(2)...	27.5	39.4	30.8	35	57	47	30	45.3	13	31	17	1	23.0
Quasqueton, Iowa.....	22.2	36.2	26.8	39	55	42	8	42.5	–2	13	–4	20	2.3
Randolph, Pa...........	24.8	36.6	27.6	42	54	42	31	43.0	0	18	12	1	14.3
Richmond, Mass.......	23.0	34.3	28.2	40	50	42	5	42.7	12	20	10	10	14.0
Sacramento, Cal.......	....	....	....	...	...	...	...	....	...	...	...	...	....
(Logan)...(3)........	....	....	....	...	...	...	...	....	...	...	...	...	....
Sacramento, Cal.......	....	....	....	...	...	...	...	....	...	...	...	...	....
(Hatch).....(4)........	....	....	....	...	...	...	24	64.1	...	...	...	16	45.5
Sag Harbor, N. Y.....	30.3	42.7	33.7	47	60	48	31	48.3	10	33	24	1	22.3
St. James, Mich.......	20.4	33.9	24.2	38	49	39	30	35.8	4	23	8	9	16.0
St. Johnsbury, Vt......	22.3	34.0	27.3	35	52	49	31	42.3	–1	20	12	1	10.3
St. Louis, Mo..........	33.6	45.5	39.1	51	70	56	5	60.8	15	24	21	20	20.0
St. Martins, C. E......	17.1	32.4	22.8	37	52	39	5	41.5	0	16	10	10	10.7
San Francisco, Cal.(5)	51.8	67.0	53.7	57	...	...	5	....	...	59	...	...	....
Saugatuck, Mich......	30.2	39.2	31.4	45	48	42	30	42.7	7	25	15	1	21.7
Savannah, Ga.........	48.9	65.1	55.1	67	84	68	15	72.7	28	45	37	1	36.9
Saybrook, Ct(6).........	30.3	38.9	31.7	37	54	42	30	43.3	20	31	19	25	27.0
Schellman Hall, Pa....	32.5	45.6	36.7	43	65	55	6	51.7	10	31	25	1	24.7
Smithsonian Inst......	37.56	46.87	40.29	52	64	54	6	56.0	22	36	25	28	33.0
Smithville, N. Y.......	25.1	31.9	28.7	40	50	45	31	45.0	10	15	10	24	16.0
Smithfield, Va.(7).....	....	....	....	...	...	...	22	42.25	...	...	...	10	16.5
Sparta, Ga............	42.4	63.2	51.8	68	82	70	16	73.0	19	43	26	1	31.0
Spencertown, N. Y....	24.9	33.4	27.6	40	57	44	31	43.7	5	22	21	1	13.7
Springdale, Ky.........	31.7	45.7	40.9	56	60	56	11	55.5	10	33	26	21	26.5
Springfield, Mass......	27.7	40.1	31.2	45	63	45	31	47.0	8	31	14	1	20.7
Steuben, Me...........	24.6	32.6	27.5	38	48	39	31	38.7	14	13	11	14	12.7
The Rock, Ga.(8)......	....	......	......	...	...	...	...	......	...	...	...	...	....
Thornbury, N. C.......	38.8	53.5	43.4	63	71	62	6	63.0	21	37	26	1	29.5
Tuscaloosa, Ala........	44.7	62.2	51.8	70	83	73	16	76.0	24	44	29	1	32.7
Upper Alton, Ill.......	30.9	44.9	36.9	48	68	58	5	58.0	12	24	15	20	17.0
Urbana, Ohio..........	27.0	40.9	32.0	45	59	48	13	46.3	–1	25	14	1	14.7
Wampsville, N. Y......	26·2	37.1	28.0	36	56	48	14	45.0	5	18	10	10	11.7
Warrington, Fa........	51.9	64.0	60.1	68	78	74	17	73.3	29	46	44	1	40.7
Westfield, Mass........	27.3	36.9	30.4	36	60	40	31	43.3	9	27	16	1	18.7
West Haverford, Pa....	31.5	43.3	....	42	66	...	31	53.0	14	30	...	1	22.0
White Marsh Island,...	....	......	......	...	...	...	...	....	...	...	...	...	....
Ga......................	48.0	62.6	53.2	64	80	66	15	69·3	30	45	33	1	36.3
Williamstown, Mass....	24.8	33.8	27.7	37	53	43	31	42.5	3	24	12	1	14.2
Winchester, Va........	34.1	48.3	37.7	54	68	55	6	55.7	16	34	20	1	24.7
Woods Hole, Mass.....	31.9	38.8	34.5	40	55	49	31	47.7	17	26	28	1	24.0
Worcester, Mass.......	28.0	38.3	31.5	42	60	64	31	53.3	11	26	17	1	21.7
Zanesville, Ohio.......	30.3	42.7	34.4	55	58	50	15	49.3	12	19	19	20	22.7

(1) Obs. at Sunrise, 9 A. M., 3 P. M. and 9 P. M., Mean at 9 A. M. 34.8, Max. 49.0, Min. 18.0.
(2) Obs. taken at Sunrise, 1 P. M. and 10 P. M.
(3) Mean for the month 54.82, Max. 76, Min. 41.
(4) Obs. taken at Sunrise, Noon, Sunset and 10 P. M. Mean for the month 56.62.
(5) Obs. taken at Sunrise, 9 A. M., Noon and 10 P. M. Mean at 9 A. M. 57.81.
(6) For the last 22 days of the month only. (7) Mean for the month 43.63, Max. 66, Min. 16.
(8) Mean for the month 51.07, Max. 72, Min. 26.

NAME OF STATION.	Mean.			Maxima.					Minima.				
							Warmest day					Coldest day.	
	7 A. M.	2 P. M.	9 P. M.	7 A.M.	2 P.M.	9 P.M.	Date	Mean temp.	7 A.M.	2 P.M.	9 P.M.	Date	Mean temp.
Alexandria, Va	47.8	62.7	54.1	66	91	77	19	77.7	27	44	34	3	36.0
All Saints, S. C	58.93	69.63	62.17	77	86	80	20	81.00	39	55	48	12	49.67
Amherst, Mass	38.3	51.3	41.9	57	76	64	25	62.0	17	23	23	2	22.3
Angelica, N. Y	38.13	52.79	39.30	62	82	62	25	66.00	16	24	14	2	21.33
Ann Arbor, Mich (Winchell)	42.7	57.5	47.0	69	84	68	18	72.6	13	27	21	2	23.2
Ann Arbor, Mich (Woodruff)	42.3	59.7	49.6	69	85	75	18	74.0	17	30	21	1	24.6
Ashland, Va	48.2	69.3	55.1	68	94	74	18	77.0	28	41	31	2	36.0
Athens, Ill	51.5	72.1	57.3	76	92	80	17	81.7	28	46	30	2	36.0
Auburn, Ala	61·4	79.5	68·6	70	96	83	19	83.0	47	62	51	7	53.7
Augusta, Ill	49.2	68.1	57.2	73	90	80	17	79.7	28	46	33	1	36.0
Austin, Texas	62.1	84.0	67.3	73	92	74	29	77.7	40	60	50	6	53.3
Baldwinsville, N. Y	38.4	47.6	43.2	56	70	64	24	63·3	15	22	18	2	20.0
Battle Creek, Mich	44.7	62.0	47.5	68	86	68	18	72.7	16	32	20	2	25.0
Bedford, Pa	43.9	61.2	49.2	75	86	74	19	74.0	20	29	28	2	25.7
Beloit, Wis	42.5	61.8	47.0	67	86	74	17	74.0	21	31	22	1	26.3
Beverly, N. Y	41.4	52.3	44.4	63	77	64	25	67·3	21	38	26	2	29.7
Bladensburg, Md(1)	50.3	66.2	50.0	66	94	76	19	79.9	28	45	34	2	38.0
Bloomfield, N. J	41.4	54.2	44.1	64	82	62	25	68.0	23	39	28	2	31.8
Boston, Mass. (2)	38.53	52.53	42.83	56	72	58	18	60.3	18	26	26	2	24.0
Brandon, Vt	36.9	50.4	40.8	53	69	60	24	60.3	12	19	22	2	18.0
Burlington, N. J	44.2	59.8	49.4	58	83	70	25	69.3	24	36	32	2	32.7
Burlington, Vt	39.7	50.6	40.6	56	76	63	24	65.0	9	19	18	2	17.0
Camden, S. C	57.6	74.9	65.1	76	97	87	19	86.7	44	53	42	3	48.0
Canton N. Y	37.5	49.7	40.8	54	72	63	24	63.0	10	22	12	2	18.0
Carmel, Me. (3)	33.5	46.8	34.7	50	70	55	24	58.3	10	21	21	2	15.5
Cedar Keys, Fa	66.2	73.0	59.0	73	81	76	28	75.0	56	62	59	1	59.0
Ceresco, Wis	45.4	57.5	43.7	73	85	68	17	72.0	18	26	16	1	21.3
Charleston, S. C	60.6	72.7	62.8	73	91	74	20	......	46	60	52	1	......
Chapel Hill, N. C	55.4	75.7	61.0	72	*	80	19	84.0	37	43	37	3	40.0
Cincinnati, Ohio. (4)	51.13	68.97	59.3	74	92	82	18	79.7	27	50	37	2	38.0
Concord N. .H	37.3	49.8	40.2	56	72	58	24	60.0	18	26	28	2	24.3
Columbus, Miss. (5)	.....	....	....	67	90	79	...	......	45	56	49		......
Cooper, Mich	49.0	60.0	48.7	70	84	70	18	72.0	26	30	22	1	28.7
Craftsbury, Vt	33.00	44.43	35.17	50	67	57	24	58.0	9	10	10	2	3.0
Crichtons Store, Va. (6)	56.3	70.9	71.7	75	93	95	19	85.0	40	50	50	3	46.7
Danville, Ky	56.4	70.9	59.1	75	89	76	18	79.3	33	53	44	2	43.3
Deaf & Dumb Inst. N. Y	43.7	52.4	46.1	60	79	65	25	68.1	27	39	28	2	32.7
Detroit, Mich	44.3	57.1	47.6	68	84	75	18	73.0	18	30	23	2	25.0
Dubuque, Iowa	49.1	63.8	53.4	71	92	80	17	77.3	28	37	31	1	32.3
Easton, Pa	42.63	59.97	46.53	56	86	65	19	68.33	22	40	30	2	29.33
Exeter, N. H	35.3	49.5	37.6	52	72	55	24	57.7	18	31	23	2	24.0
Flint, Mich	43.0	60.0	46.6	71	87	71	...	....	14	27	20	...	....
Ft. Madison, Iowa. (7)	46.5	66.8	58.8	73	90	82	25	80.3	26	41	38	2	35.0
Frederick, Md	47.7	63.2	52.5	68	88	72	19	76.2	27	43	32	2	34.9
Friendship, Tenn (8)	63.1	81.1	66.2	74	91	79	19	80.7	52	66	57	15	62.7
Gallipolis, Ohio	53.30	70.15	54.23	75	91	74	19	78.00	29	45	29	2	34.33
Gardiner, Me	38.3	48.6	39.9	51	59	55	24	56.8	18	26	22	2	22.0
Garlandsville, Miss. (9)	63.5	84.4	71.8	76	98	80	19	84.0	45	45	52	1	59.0
Germantown, Ohio	49.3	65.9	53.4	64	90	70	18	77.1	21	40	27	2	29.7
Gettysburg, Pa	45.5	61.9	49.3	72	89	70	19	77.0	24	39	28	2	31.0
Glenwood, Tenn	57.8	71.1	60.3	73	90	80	19	80.3	36	46	43	2	43.2
Gouverneur, N. Y	38·1	51.1	41.6	60	70	60	17	62.0	10	26	20	2	18·7
Grand Rapids, Mich	41.3	56.9	46.8	76	78	68	18	74.0	16	30	18	1	24.7

(1) Incomplete.

*102

(2) Obs. taken at 6 A. M. 2 P. M. and 10 P. M.

(3) No obs. on the second at 9. P. M.

(4) Obs. taken at 6 A. M. 1 P. M. and 9 P. M.

(5) Mean for the month, 69.07.

(6) Obs. irregular, at 8 A. M. Noon, and 5 P. M. No obs. on the nineteenth at noon.

(7) Obs at 6 A. M. noon and7 P. M.

(8) No obs. for the first 13 days.

(9) Obs. at 7 A. M. 1 P. M. anb 9 P. M.

NAME OF STATION.	Mean.			Maxima.					Minima.				
	7 A. M.	2 P. M.	6 P. M.	7 A.M	2 P.M	9 P.M	Warmest day Date	Warmest day Mean temp.	7 A.M	2 P.M	9 P. M	Coldest day. Date	Coldest day. Mean temp.
Granville, Ohio......	49.5	64.4	53.5	72	76	75	18	77.0	22	40	30	2	32.3
Great Falls, N. H...	36.80	50.63	38.87	50	95	58	24	60.00	14	20	22	2	19.00
Green Springs, Ala.(1)	67.4	80.9	69.4	80	06	78	18	82.7	23	61	50	7	55.0
Hamilton, Ca..(2)...	45.80	....	45.16	65	...	63	24	60.33	28	...	22	2	24.00
Harrisburg, Pa.....	46.8	60.3	53.7	64	83	73	19	72.0	22	41	32	2	34.7
Hillsboro, Ohio.....	49.6	64.0	54.3	71	87	75	18	77.5	48	39	33	2	31.3
Jackson, Ohio (Crookham).......	49.7	69.7	52.5	73	94	70	19	76.7	28	42	27	2	33.3
Jackson, Ohio (Gilmor)............	49.9	69.4	54.1	78	93	76	19	78.3	30	42	30	2	35.7
Jackson, Ohio (Wood)......(3)....	52.2	72.2	58.2	78	93	76	19	78.3	32	54	35	11	43.0
Jacksonville, Fa.....	67.13	77.26	66.53	74	89	81	20	80.6	59	61	53	7	58.33
Key West, Fa........	71.0	78.6	......	77	84	...	6	79.57	63	70	...	8	65.5
Knox Hill, Fa..(4)...	62.9	81.4	66.9	71	91	79	20	79.7	45	32	24	2	19.0
Knoxville, Tenn.....	54.2	70.9	59.3	75	92	77	18	78.8	36	46	38	7	43.00
Lacqiparle, Minn...	45.8	61.3	48.6	67	90	68	17	74.0	24	30	23	9	28.0
Lewisburg, Va......	46.5	69.1	52.7	58	92	65	18	70.3	29	47	38	2	38.3
Lima, Pa..............	45.2	58.0	48.0	63	85	73	19	72.6	26	39	28	3	32.6
Lodi, N. Y...........	42.2	51.1	41.8	64	78	64	24	66.3	12	22	14	2	19.7
Londonderry, N. H.	37.6	51.2	42.0	57	78	60	24	61.0	16	23	25	2	21.3
Lowville, N. Y.(5)...	33.1	47.4	36.5	48	70	60	24	57.3	9	20	14	2	16.3
Madison, Ind........	50.4	65.5	55.3	73	89	75	18	75.7	29	41	37	2	35.8
Manchester, N. H(6	34.8	49.1	42.2	51	64	56	22	53.3	15	28	30	2	24.3
Manchester, Ill..(7).	50.6	70.8	56.6	72	94	78	18	80.7	30	41	33	2	36.7
Marietta, Ohio......	44.9	65.5	51.5	70	87	75	18	72.3	20	39	27	2	30.0
Mendon, Mass.......	40.1	51.4	41.4	53	71	54	24	57.3	17	24	25	2	22.0
Millersburg, Ky.....	50.7	66.8	59.5	75	84	80	18	79.7	26	42	38	2	35.3
Milton, Ind..........	47.9	63.4	52.0	76	85	74	18	78.5	22	38	29	2	30.5
Milwaukie, Wis......	44.3	54.2	41.6	69	84	62	18	68.3	24	27	20	1	24.3
Morrisville, Pa.	42.5	57.6	47.1	54	82	67	19	67.7	26	41	30	2	34.7
Moss Grove, Pa.....	41.62	55.20	45.58	68	84	68	18	72.0	14	22	14	2	17.33
Mount Vernon, O...	48.7	67.2	50.1	69	90	74	24	75.00	20	38	24	2	27.7
Muscatine, Iowa....	45.8	65.9	51.2	72	87	78	25	77.3	24	44	25	10	32·7
Nantucket, Mass....	43.3	49.0	42.0	50	63	49	19	53.2	28	34	30	2	31.5
Newark, Ohio.......	47.9	62.8	51.5	74	88	73	18	76.7	22	36	27	2	28.3
Newark, N. J..(8)...	....	....	....	...	...	...	25	71.4	...	...	...	2	33.4
New Bedford, Mass.	40.6	51.3	41.6	58	76	50	19	60.3	20	32	25	2	27.2
Newburyport, Mass.	38.4	50.7	41.7	54	77	59	24	61.4	17	24	26	2	23.0
New Harmony, Ind	52.73	68·73	57.60	74	89	79	18	80.33	31	46	38	2	39.00
New Lisbon, Ohio...	44.53	57.57	47.03	70	81	67	18	71.0	18	35	26	2	27.33
New London, Ct.....	43.8	54.0	43.3	54	80	56	19	62.0	24	34	28	2	29.3
New Orleans, La....	63.3	78.8	69.7	72	85	78	22	77.8	44	67	58	1	57.2
New Wied, Texas,	66.6	84.6	68.8	74	94	76	19	81.3	32	65	47	6	55.3
Norristown, Pa......	44.9	55.8	48.4	60	82	70	25	69.7	30	39	33	1	34.7
North Attleboro, Mass................	40.4	50.8	41.2	56	71	59	18	58.3	18	25	23	2	23.5
Oberlin, Ohio..(9)...	48.7	60.4	48.5	72	86	70	18	75.3	19	32	24	2	19.0
Oldtown, Me.........	30.7	44.2	34.6	46	62	50	25	51.3	10	17	14	2	13.7
Ottowa, Ill............	47.8	67.0	50.6	70	91	77	17	78.7	23	39	23	1	31.3
Oxford, Miss.	63.3	75.8	65.7	80	86	80	18	80.7	46	54	47	6	50.5
Pella, Iowa..........	46.5	67.0	52.0	85	92	78	24	84.0	26	40	26	1	32.0
Penn Yan, N. Y.(10)	36.4	53.6	45.2	59	79	71	18	68.3	15	25	18	2	21.3
Perry, Me...... (11).	32.0	45.1	36.1	42	62	46	18	49.7	21	23	19	2	21.3
Perrysburg, Ohio (7)	53.7	64.8	53.3	80	89	77	18	82.0	28	40	26	1	31.3

(1) Obs. irregular. (2) Obs. taken at 9 A. M. and 9 P. M. (3) No obs. before the sixth.
(4) Obs. omitted from the tenth to the eighteenth. (5) Obs. at Sunrise, noon and 9 P. M.
(6) Obs. very irregular from the 17th to the 26th. (7) Obs. at 7 A. M. 1 P. M. and 9 P. M.
(8) Mean for the month 48.95, Max. 85, Min. 27.25. (9) Only one obs. on the 2nd, at 7 A. M.
(10) Obs. taken at Sunrise, 2 P. M. and Sunset. (11) Obs. at Sunrise, 2 P. M. and 9 P. M

NAME OF STATION.	Mean.			Maxima.					Minima.				
							Warmest day					Coldest day.	
	7 A. M.	2 P. M.	9 P. M.	7 A.M	2 P.M	9 P.M	Date	Mean temp.	7 A.M	2 P.M	9 P.M	Date	Mean temp.
Pittsburgh, Pa......	......	......	......	...	...	...	...	......	...	...	...	...	......
(Oakland Station).	44.47	61.23	47.73	71	83	70	18	72.00	19	29	24	2	24.00
Pittsburg, Pa..(Ma-	......	......	......	...	...	...	...	......	...	...	...	...	......
rine Hospital)..[1].	48.1	58.5	54.1	66	83	78	25	69.5	33	33	31	2	32.5
Philadelphia, Pa...	47.0	59.6	52.1	63	86	75	25	74.0	26	38	32	2	34.0
Plattsville, Wis......	46.5	65.0	54.5	76	88	78	17	76.7	27	35	26	1	29.3
Pocopson, Pa.........	44.4	60.2	49.0	62	86	74	19	72.7	26	39	29	2	33.3
Pomfret, Conn......	38.2	50.1	39.9	56	77	53	25	59.3	16	26	22	2	22.0
Pottsville, Pa.......	39.5	56.8	44.1	58	85	64	19	69.2	25	35	25	2	29.7
Poultney, Iowa.....	44.5	63.2	48.1	69	92	74	17	78.3	25	35	24	1	28.0
Princeton, Mass....	36.2	46.6	38.7	54	67	55	24	55.7	13	19	20	2	17.8
Providence, R. I.[2]	38.7	52.5	41·0	54	76	54	18	58.0	18	30	24	2	25.0
Quasqueton, Iowa...	46.1	63.9	52.4	73	92	75	17	77.7	25	37	24	1	28.7
Randolph, Pa........	42.7	59.8	44.3	68	88	70	18	74.0	16	24	16	2	20.3
Richmond, Mass.....	34.9	48.7	40.2	58	72	60	18	56.7	10	22	18	2	16.7
Sag Harbor, N. Y..	41.6	54.9	43.6	55	85	64	25	64.3	24	36	32	2	30.7
St. James, Mich.....	37.86	48.81	36.80	60	73	63	29	61.67	20	28	15	1	21.16
St. Johnsbury, Vt..	34.70	47.14	38.70	51	70	63	24	56.33	12	19	19	2	17.00
St. Louis,Mo.........	55.72	70.13	60.80	75	93	80	25	82.20	32	47	42	2	40.33
St. Martins, Ca. E..	32.2	47.6	37.6	49	71	54	24	57.3	5	15	11	2	15.1
Saugatuck, Mich...	46.2	59.3	47.2	73	81	73	17	72.3	23	32	21	1	27.3
Savannah, Ga.......	60.5	76.6	64.5	77	99	83	20	85.9	47	62	53	12	55.9
Savannah, Ohio[3]..	55.5	66.8	55.0	80	93	84	18	85.7	30	38	36	6	36.0
Saybrook, Conn[4]..	39.1	48.4	40.9	50	64	49	25	54.0	19	34	28	2	27.7
Schellman Hall Md.	48.3	63.5	50.1	72	87	75	19	77.3	28	43	30	2	34.3
Smithsonian Inst...	48.4	63.9	54.9	65	97	80	25	77.6	30	45	34	3	38.3
Smithville, N. Y.....	39.0	49.2	40.1	57	70	65	24	64.0	14	25	20	2	19.7
Smithfield, Va...[5]..	......	....	....	...	...	...	19	78.5	...	...	...	3	40.5
Sparta, Ga..........	55.8	78.5	64.5	72	98	86	18	82.7	39	54	45	3	49.0
Spencertown, N. Y.	39.8	49.2	39.6	64	72	52	25	60.3	20	24	20	2	24.0
Springfield, Mass...	38.5	54.0	42.8	56	80	60	25	63.3	16	26	20	2	22.3
Steuben, Me.........	35.2	40.3	35.7	51	59	45	16	46.3	17	22	18	2	19.0
The Rock, Ga. [6]..	......	......	......	...	...	...	...	......	...	...	...	...	......
Thornbury, N. C....	51.9	72.3	57.0	68	79	75	26	76.6	34	37	40	2	38.8
Unionville, Ohio....	47.9	56.7	46.5	80	85	68	18	77.7	26	30	26	2	27.7
Upper Alton, Ill....	51 7	69.0	59.8	77	89	81	25	81.3	28	44	38	2	38.7
Wampsville, N. Y...	40.23	53.20	40·53	64	78	56	24	65.0	16	25	22	2	22.0
Warrington, Fa..[7]	64.3	74.1	70.8	74	80	78	29	76.7	50	66	58	2	61.3
Westfield, Mass.....	37.23	50.70	42.67	53	76	65	25	60.67	16	24	23	2	22.0
W Haverford, Pa.[8]	44.4	59.0	......	62	85	...	19	71.0	23	38	...	2	30.5
Whitemarsh Is, Ga..	61.1	73.1	62·8	74	93	76	20	81.0	48	61	49	7	54.0
Williamstown,.......	......	......	....	...	...	...	...	......	...	...	...	...	......
Mass................	37.2	48.7	41.7	53	67	60	19	57.2	13	21	22	2	18.7
Winchester, Va......	50.6	67.0	53.6	78	90	78	19	79.3	28	42	32	2	34.7
Wood's Hole, Mass..	41.2	49·5	43.5	52	63	58	19	56.67	21	30	29	2	27.3
Worcester, Mass....	38.9	50.2	41.3	58	67	58	18	58.7	20	26	24	2	24.0
Zanesville, Ohio·...	47.7	65.4	53.9	70	90	72	25	76.7	24	45	31	2	35.0

(1) Obs. at Sunrise, 9 A. M. 3 P. M. and 9 P. M. At 9 A. M. as follows, Mean 49.8 Max. 75 Min 28.
(2) Obs. taken at 6 A. M. 1 P. M. and 10 P. M.
(3) No obs. before the fifth.
(4) Obs. at 7 A. M. 2 P. M. and 11 P. M.
(5) Mean for the month 29.51, Max. 92, Min. 22.
(6) Mean for the month 62.70, Max. 76, Min. 48.—Obs. taken morning, 3 P. M. and evening.
(7) Obs. at Sunrise, Noon, and Sunset.
(8) No obs. at 9 P. M. Obs. after the thirteenth made at Philadelphia.

NAME OF STATION.	Mean.			Maxima.					Minima.				
	7 A. M.	2 P. M.	9 P. M.	7 A.M	2 P.M.	9 P.M	Warmest day Date	Warmest day Mean temp.	7 A.M	2 P.M	9 P.M	Coldest day. Date	Coldest day. Mean. temp.
Alexandria, Va	58.8	71.7	61.8	71	85	77	16	77.7	45	52	44	9	48.0
All Saints, S. C	66.9	75.6	69.3	76	83	77	24	78.0	53	65	55	10	57.7
Amherst, Mass	50.2	66.0	53.7	68	81	66	16	70.3	41	42	38	9	40.8
Angelica, N. Y	49.68	63.94	49.39	70	86	66	31	72.00	36	38	32	8	36.6
Ann Arbor, Mich. (Winchell)	55.0	70.3	53.8	67	83	69	23	72.4	35	33	33	7	35.7
Ann Arbor, Mich. (Woodruff)	51.4	69.2	57.2	69	85	73	23	75.3	34	35	33	7	36.7
Ashland, Va	57.1	75.2	61.6	72	90	76	24	77.3	41	48	45	8	44.7
Athens, Ill	59.0	74.2	61.5	76	96	83	22	85.0	38	56	39	8	46.0
Auburn, Ala...(1)	68.9	84.3	72.3	87	99	89	26	92.5	52	68	61	9	61.3
Augusta, Ill	57.0	70.7	60.8	77	93	84	22	84.0	38	60	46	8	48.0
Austin, Texas	68.6	87.0	76.1	76	92	82	24	83.0	60	76	64	10	68.0
Baldwinsville, N. Y	48.8	60.3	53.5	65	76	71	31	70.7	40	38	38	8	38.7
Battle Creek, Mich	55.1	71.9	55.7	70	94	68	22	77.3	35	36	33	7	38.3
Bedford, Pa	54.1	71.6	60.3	71	85	73	24	73.3	35	47	37	8	40.0
Beloit, Wis	51.8	73.2	55.1	69	93	81	22	81.0	34	49	34	8	39.0
Beverly, N. Y	52.7	63.6	57.1	73	74	71	16	72.0	40	38	38	9	38.7
Bloomfield, N. J	53.0	69.1	55.6	68	85	71	16	73.3	40	41	39	9	40.0
Boston, Mass...(2)	47.87	64.29	52.61	65	82	66	16	73.5	36	46	42	3	44.0
Brandon, Vt	49.0	64.0	53.9	64	82	71	31	71.7	37	47	43	9	44.3
Burlington, Vt	51.3	65.8	54.8	67	85	71	31	74.3	40	50	43	7	46.0
Burlington, N. J	55.0	70.1	59.9	70	85	78	16	77.7	40	48	41	9	43.0
Camden, S. C	66.0	79.7	69.1	76	93	84	22	83.3	52	66	57	9	59 7
Canton, N. Y	51.4	68.5	55.0	70	88	74	31	77.3	36	53	42	7	44.3
Carmel, Me. (3)	47.3	64.4	47.2	60	86	70	31	70.0	35	44	36	20	38.3
Cedar Keys, Fa...(4)	72.3	78.1	74.4	81	84	81	27	82.5	60	73	67	11	68.3
Ceresco, Wis	58.8	68.8	52.1	73	90	69	22	73.0	36	44	32	7	38.0
Charleston, S. C	68.9	79.3	80.0	77	90	81	23	......	55	69	60	10	.....
Chapel Hill, N. C	60.3	77.3	65·6	70	91	78	24	78.6	45	57	51	9	54.0
Cincinnati, Ohio..(5)	57.1	75.0	64.3	71	91	78	23	79.0	37	51	46	8	44.7
Cleveland, Ohio..(6)	55.1	63.9	55.8	72	86	74	30	76.3	36	37	36	8	37.7
Columbus, Miss...(7)	......	......	......	76	94	87	...	......	48	70	60	..	.....
Concord, N. H	50.2	64.5	52.4	66	80	69	15	68.3	40	47	42	9	43.7
Cooper, Mich	59.1	70.3	56.9	80	92	68	22	80.0	33	36	36	7	38.7
Craftsbury, Vt	44.81	60.26	49.39	60	80	67	31	69.00	32	45	38	7	39.0
Crichton's Store, Va. (8)	64.1	77.8	76.0	73	90	89	1	83.7	52	60	64	4	58.7
Deaf & Dumb Inst. N. Y	55.6	65.4	57.0	69	81	71	16	74.0	38	41	39	9	39.4
Detroit, Mich	53.7	66.6	54.3	68	82	71	30	73.7	37	34	32	7	36.3
Dubuque, Iowa	57.7	72.7	61.3	69	99	79	22	81.0	39	53	45	8	48.0
Easton, Pa.	53.71	72.45	59.52	69	88	74	24	73.67	40	45	39	8	44.0
Exeter, N. H	47.1	60.6	48.5	68	80	65	31	68.7	36	45	39	8	41.3
Flint, Mich	53.3	71.3	54.6	70	84	69	...	......	34	30	36	..	......
Fort Madison, Iowa, (9)	56.7	72.5	65.0	78	96	92	22	88.7	36	55	47	8	47.0
Frederick City, Md	59.0	72.8	61.7	75	87	77	16	78.2	45	52	46	9	47.9
Friendship, Tenn	63.6	77.6	65.5	76	96	84	22	83.0	42	50	45	8	50.0
Gallipolis, Ohio	59.5	76.03	66.0	72	88	82	1	78.0	45	59	44	8	49.7
Gardiner, Me	51.0	61.3	50.9	62	79	63	31	68.0	41	45	42	21	43.0
Garlandsville, Miss	75.2	92.0	78.9	90	*	90	23	94.3	56	70	62	9	62.7
Germantown, Ohio	57.5	72.5	57.5	68	90	70	23	74.3	37	47	38	8	42.8

(1) No obs. on the 26th at 9 P. M.

(2) Obs. taken at 6 A. M. 2 P. M. and 10 P. M.

(3) No obs. before the 15th.

(4) No obs. after the 27th, None on the 27th at 9 P. M.

(5) Obs. taken at 6 A. M. 1 P. M. and 9 P. M.

(6) Several of the morning obs. taken at Sunrise

(7) Mean for the month 75.42.

(8) Obs. taken at 7 A. M. Noon and 6 P. M. Incomplete.

(9) Obs. taken at 6 A. M. Noon and 7 P. M.

*107

NAME OE STATION.	Mean.			Maxima.					Minima.				
	7 A. M.	2 P. M.	9 P. M.	2 P.M	9 P. M	7 A.M	Warmest day Date	Warmest day Mean temp.	7 A M	2 P.M	9 P.M	Coldest day. Date	Coldest day. Mean. temp
Gettysburg, Pa.........	57.9	73.8	59.8	71	88	78	24	76.3	40	49	40	8	43.3
Glenwood, Tenn.......	58·9	75.1	64.2	73	92	84	22	82.8	42	58	47	8	50.3
Gouverneur, N. Y....	53.6	65.3	56.7	68	80	70	31	72.7	40	46	40	9	44.7
Grand Rapids, Mich....	51.3	68.5	55.9	69	92	77	22	77.0	35	39	37	7	38.7
Granville, Ohio.........	54.6	71.3	58.4	69	86	76	22	75.7	37	44	39	8	40.7
Great Falls, N. H......	49.23	64.24	49.73	67	81	69	31	69.3	40	45	38	9	42.8
Green Springs, Ala....	73.1	91.1	76.5	90	*	90	26	93.0	53	74	63	9	63.7
Hamilton, Ca...(1)......	58.0	......	55.9	73	...	72	31	72.5	38	...	37	8	39.5
Harrisburg, Pa..	57.5	72.9	64.4	72	87	79	24	78.0	44	50	49	8	48.7
Hillsborough, Ohio....	56.3	70.9	59.6	66	88	74	23	73.7	34	45	40	8	40.7
Jackson, Ohio, ...	....	....	....	...	...	...	...	....	...	...	...	...	....
(Gilmor)...............	58.8	77.4	62.5	74	93	84	23	81.3	36	50	43	8	45.0
Jackson, Ohio...........	....	....	....	...	...	...	...	......	...	...	...	...	....
(Crookham)...........	58.4	76.7	60.9	73	88	80	22	80.0	40	47	44	8	43.7
Jacksonville, Fa......	72.48	83.54	73.35	83	†	89	23	91.00	62	71	61	10	65.7
Key West, Fa.........	73.3	82.2	......	80	88	...	29	83.5	71	71	...	16	70.50
Knox Hill, Fa.	70.0	83.9	71.6	79	98	79	23	83.3	57	58	62	9	59.7
Knoxville, Tenn........	61.6	76.2	64.6	76	90	78	23	81.3	44	58	46	8	50.7
Lac qui parle, Min.....	56.5	70.3	55.8	69	83	68	21	70.7	40	54	43	23	48.7
Lewisburg, Va.........	57.0	76.8	59.2	70	92	70	24	77.0	40	50	42	8	44.3
Lima, Pa.	56.0	68.0	57.5	70	83	73	16	75.3	41	46	43	9	44.6
Lodi, N. Y................	51.9	62.7	52.6	69	79	69	31	72.0	34	41	34	8	36.3
Lowville, N. Y. ..(2) ...	44.06	64.94	51.26	62	84	70	31	66.0	30	43	33	9	38.25
Madison, Wis...........	55.2	67.9	61.3	74	87	78	22	79.7	36	45	51	8	47.7
Madison, Ind............	58.2	72.5	62.6	73	89	79	23	79.0	39	51	43	8	44.7
Madrid, N. Y...(3)......	59.55	70.85	59·05	62	78	72	31	69.8	56	58	51	27	59.5
Manchester, Ill.(4)......	59.2	74·2	60.0	79	94	80	22	84.3	36	58	46	8	46.7
Manchester, N. H......	48·96	66.67	54.52	67	81	70	31	70.33	38	48	40	10	44.67
Marietta, Ohio.........	55.5	75.0	58.3	69	85	71	24	73.7	38	53	43	8	46.0
Meadville, Pa...........	52.6	67.5	55.2	69	86	73	30	75.0	36	40	34	8	37.0
Mendon, Mass..........	51.1	62.0	51.2	68	76	67	31	68.1	36	40	39	9	40.0
Millersburg, Ky........	61.9	73.3	66.8	78	86	80	23	80.7	36	48	40	8	41.3
Milton, Ind..............	56.8	71.9	58.7	74	93	75	22	79.2	38	46	39	8	42.2
Milwaukie, Wis........	50.8	61.1	49.1	68	82	68	15	71.0	38	38	35	7	38.0
Morrisville, Pa.	53.9	68.6	56.6	68	83	72	16	73.3	38	44	40	9	40.7
Moss Grove, Pa.........	54.23	66.03	53.23	68	87	70	30	72.0	32	42	33	8	35.7
Muscatine, Iowa.......	53.7	70.1	57.5	78	86	76	21	75·7	27	52	42	9	42·7
Nantucket, Mass.......	51.2	57.0	48.7	60	67	59	30	61·7	41	41	39	10	42.8
Newark, N. J. (5)......	......	......	......	...	...	...	16	74.5	...	...	...	9	43.0
New Bedford, Mass....	50.9	60.8	49.6	62	73	59	25	63.7	41	38	37	9	39.0
Newburyport, Mass....	50.3	62.0	51.9	71	82	68	16	71.7	40	44	41	2	41.9
New Harmony, Ind.....	60.97	75.32	63.55	79	91	86	22	85.3	42	59	46	8	49.3
New Lisbon, Ohio.....	55.16	67.77	56.71	69	80	70	23	71.67	38	39	38	7	51.0
New London, Conn.....	54.6	65.4	52·3	65	79	61	25	98.0	43	45	36	9	41.3
New Orleans, La.......	70·9	83.7	75.8	78	93	83	28	84.0	63	71	67	10	67.5
New Wied, Texas......	75.4	93·4	75.6	83	‡	86	25	88.3	63	78	67	8	71.7
Norristown, Pa.........	52.1	66.9	56.9	65	83	73	16	72.0	40	46	42	9	43.3
North Attleboro,......	....	....	....	...	...	...	...	....	...	...	...	...	....
Mass....................	50.7	62 8	49.3	66	77	60	16	67.8	39	39	36	9	39.0
Oberlin, Ohio..........	55.0	68.3	54.6	71	90	70	23	76.3	38	37	36	8	38.0
Oldtown, Me.	42.8	57·3	44.6	52	74	58	31	61.3	33	42	35	1	39·7
Oswego, N. Y..........	49.0	61.6	51.6	65	83	72	31	68.0	37	39	41	8	40.3
Ottawa, Ill..............	59.7	73.3	56.5	80	96	78	22	84.7	41	50	36	8	44.3
Oxford, Miss...........	74.6	80.7	70.0	92	91	82	22	86.3	53	66	51	9	58.0

(1) Obs. taken at 9 A. M. and 9 P. M. *103 †101 ‡103
(2) Obs. taken at Sunrise, Noon and 9 P. M.
(3) Incmplete.
(4) Obs. taken at 7 A. M. 1 P. M. and 9 P. M.
(5) Obs. taken at 7 A. M and 5 P. M. Mean for the month 59.09, Max. 83, Min. 32.75

NAME OF STATION.	Mean.			Maxima.					Minima.				
	7 A. M.	2 P. M.	9 P. M.	7 A.M	2 P.M	9 P.M	Warmest day Date	Warmest day Mean temp.	7 A.M	2 P. M	9 P.M	Coldest day. Date	Coldest day. Mean. temp.
Pella, Iowa	53.6	72.4	57.4	71	91	72	21	74.3	28	46	40	8	42.7
Penn Yan, N. Y.(1)	44.5	63.2	54.4	60	81	69	31	67.10	33	36	36	8	36.67
Pittsburg, Pa. (Oak Station)	54.52	66.39	26.65	70	81	79	23	71.3	36	39	38	7	44.0
Pittsburg, Pa. (Marine Hospital,) (2)	54.7	68.0	63.3	72	80	75	24	71.5	38	44	41	8	48.2
Philadelphia, Pa	57.8	71.1	62.4	70	87	78	16	77.5	42	50	44	9	45.3
Perry, Me(3)	39.8	52.7	43.7	50	68	57	25	56.3	33	38	34	1	37.0
Perrysburg, Ohio(4)	60.8	70.2	58.7	77	84	75	23	76.7	38	40	35	7	41.7
Platteville, Wis	57.6	75.8	63.7	72	98	81	22	83.7	34	60	44	8	48·0
Pocopson, Pa	55.6	71.5	58.9	72	86	76	16	78.0	42	47	43	9	45.2
Pomfret, Conn	49.7	62.7	50.3	66	78	62	16	68.7	40	40	35	9	38.3
Pottsville, Pa	50·06	69.80	53.22	72	85	73	24	73.0	36	43	39	9	41.00
Poultney, Iowa	56.6	73.1	54.0	75	90	70	22	78.0	40	52	34	7	44.7
Princeton, Mass	47.9	60.2	50.5	65	77	67	16	66.2	36	39	35	9	36.8
Providence, R. I (5)	49.2	63.5	50.3	65	78	64	16	68.3	38	38	35	9	38.0
Quasqueton, Iowa	54.6	72.2	56.5	71	88	72	21	75.8	35	54	34	8	46.0
Randolph, Pa	53.1	68.2	53.3	68	86	70	30	73.7	32	37	32	8	33.7
Richmond, Mass	46.5	65.5	55.5	64	82	70	15	71.0	28	38	36	4	34.3
Sag Harbor, N. Y	52.7	67.3	52.2	65	85	61	15	69.7	43	39	37	9	39.7
St. James, Mich	50.4	60.9	47.1	68	75	69	15	59.67	35	45	38	2	39.67
St. Johnsbury, Vt	51.04	62.48	51.93	65	83	69	31	71.33	40	41	42	21	46.33
St. Louis, Mo	62.37	73.55	64·37	79	92	81	22	84.02	45	54	49	8	51.02
St. Martins, Ca. E	46.9	67.2	52.0	62	89	73	31	74.9	32	52	36	7	42.9
Saugatuck, Mich	58.4	66.8	56.7	74	85	72	30	76.7	43	47	41	7	43.7
Savannah, Ga	68.7	82.6	70.9	82	95	82	23	85.8	59	71	60	10	63.7
Savannah, Ohio	56.3	69.9	55.0	72	94	72	29	77.0	34	38	35	8	37.0
Saybrook, Conn.(6)	51.1	61.3	50.6	62	74	60	25	68.0	40	40	38	9	40.7
Schellman Hall, Md	58.8	73.5	59.5	71	85	78	16	77.7	43	53	43	9	47.0
Smithsonian Inst	57.2	72.3	63.2	70	85	77	24	76.7	43	53	47	9	48.2
Smithville, N. Y	48.5	63.7	50.8	68	84	76	31	74.7	32	42	37	9	41.7
Smithfield, Va (7)	......	......	......	...	...	...	24	76.75	...	...	...	8	50.5
Sparta, Ga	61.3	85.2	68.5	75	*	84	23	86.3	47	68	53	9	56.0
Spencertown, N. Y	50.6	63.1	47.6	72	84	66	31	69.0	35	37	32	9	37.0
Springdale, Ky	53.05	75.66	66.60	69	95	84	23	81.5	34	55	55	8	46.0
Springfield, Mass	50.8	66.9	55.5	68	82	67	16	71.3	40	42	38	9	40.7
Steuben, Me	43.5	52.7	45.8	55	73	58	31	62.0	33	33	33	1	35.3
The Rock, Ga. (8)	....	....	....	...	...	...	...	....	...	...	...	...	....
Unionville, Ohio	55.8	62.5	54.4	70	82	72	30	73.3	35	37	34	8	37.3
Upper Alton, Ill	59.1	72.4	63.3	73	88	81	22	80.0	38	54	46	8	47.3
Urbana, Ohio	58.4	74.5	59.3	76	91	80	29	79.7	33	45	35	8	40.0
Wampsville, N. Y	50.22	67.58	51.26	68	86	72	31	75.3	38	42	37	8	41.0
Warrington, Fa. (9)	72.1	78.2	77.1	78	84	82	29	81.0	60	70	70	10	67.7
Westfield, Mass	45.50	65.65	53.48	71	82	66	31	67.3	38	43	37	10	47.0
W. Haverford, Pa(10)	53 6	69.9	......	68	84	...	16	75.5	42	43	...	9	42.5
Whitemarsh Is, Ga	70.2	79.1	71.0	83	90	81	23	83.0	60	67	61	10	62.7
Williamstown, Mass(11)	48.89	61.08	52.70	71	80	69	31	71.33	36	42	38	10	44.70
Winchester, Va	60.7	74.2	61.6	70	90	74	24	76.7	45	52	46	8	48.7
Worcester, Mass	51.0	63.4	15.2	67	79	67	16	70.3	35	41	39	9	92.3

(1) Obs. taken at Sunrise, 2 A. M. and Sunset. *100

(2) Obs. taken at Sunrise, 9 A. M. 3 P. M. and 9 P. M. Mean at 9 A. M. 59.8, Max. 72, Min. 43.

(3) Obs. taken at Sunrise 2 P. M. and 9 P. M.

(4) Obs. taken at 7 A. M. 1 P. M. aud 9 P. M.

(5) Obs. taken at 6 A. M. 1 P. M. and 10 P. M.

(6) Obs. at 7 A. M. 2 P. M. and 11 P. M. — None on the 25th at 11 P. M.

(7) Mean for the month 63.58, Max. 86, Min. 60.

(8) Mean for the month 69.79, Max. 82, Min. 56.—Obs. taken morning, 3 P. M. and evening

(9) Obs. taken at Sunrise, Noon and Sunset.

(10) Obs. until the 9th P. M. made at Philadelphia.

(11) Hours of observation variable.

NAME OF STATION.	Mean.			Maxima.					Minima.				
	7 M. A.	2 P. M.	9 P. M.	7 A.M	2 P.M.	9 P.M.	Warmest day Date	Warmest day Mean temp.	7 A. M	2 P. M	9 P. M	Coldest day. Date	Coldest day. Mean temp.
Alexandria, Va	67.45	77.33	68.33	86	94	85	29	87.33	55	56	57	4	58.33
All Saints, S. C	72.97	75.33	74.90	81	83	79	23	80.00	61	68	63	4	64.00
Amherst, Mass	60.65	71.55	62.31	82	92	81	30	84.67	48	57	53	12	53.67
Angelica, N. Y	57.27	68.33	56.83	73	93	73	30	79 00	40	51	42	4	45.67
Ann Arbor, Mich (Woodruff)	58.07	70.06	61.07	77	91	80	28	81.33	40	52	44	3	44.67
Ann Arbor, Mich (Winchell)	59.99	68.46	67.83	81	91	73	29	81.03	42	48	43	3	44.07
Ashland, Va	63.97	76.37	65.27	77	93	88	29	82.33	49	56	49	3	51.33
Athens, Ill	65.30	78.53	66.00	80	94	80	28	84.67	50	60	43	2	52.33
Auburn, Ala	70.11	83.90	74.60	82	93	82	28	84.33	59	70	59	3	64.00
Augusta, Ill	63.60	76.57	67.83	78	90	88	28	82.67	49	55	50	3	54.33
Austin, Texas	72.17	85.57	76.67	79	92	82	29	83.33	63	72	66	7	69.00
Baldwinsville, N. Y	56.50	66.50	60.73	74	89	80	30	81.00	44	50	46	12	47.33
Battle Creek, Mich	59.70	78.83	59.90	84	95	79	29	86.00	42	47	41	3	48.33
Bedford, Pa	60.17	74.45	63.31	77	93	80	29	83.00	45	61	50	18	52.33
Beloit, Wis	61.93	72.20	61.43	74	92	79	28	80.33	41	53	43	3	48.33
Beverly, N. Y	63.00	70.04	64.80	80	89	80	30	83.00	53	62	53	5	56.33
Bladensburg, Md.(1)	66.36	79.72	64.82	83	98	80	24	79.00	57	62	55	18	59.67
Bloomfield, N. J	62.57	74.83	64.00	80	97	83	30	85.33	51	64	53	13	57.83
Boston, Mass(2)	59.37	74.67	63.73	75	92	83	29	82.33	44	60	56	13	53.33
Brandon, Vt	61.20	66.37	60.60	79	90	80	30	83.16	49	52	50	8	51.00
Burlington, Vt	60.97	71.70	61.00	80	92	82	30	84.60	51	58	49	12	54.67
Burlington, N. J	66.33	76.30	68.87	80	93	88	29	86.93	55	59	56	19	60.00
Camden, S. C	72.03	82.96	71.87	79	90	82	30	83.67	61	71	61	4	64.67
Canton, N. Y	60.07	68.67	61.17	74	91	80	30	81.67	50	52	49	12	50.67
Carmel, Me	57.15	72.28	54.77	74	92	70	29	78.00	49	50	40	9	50.67
Cedar Keys, Fa.(3)	78.50	84.63	79.13	82	86	82	23	83.33	71	82	75	15	77.00
Chapel Hill, N. C	68.13	83.30	70.90	77	99	84	30	86.67	53	69	57	4	59.67
Charleston, S. C	73.43	81.73	74.20	80	90	80	20	......	60	72	67	4	......
Chestertown, Md(4)	72.15	81.75	71.95	82	93	80	29	85.00	58	62	56	12	60.00
Cincinnati, Ohio (5)	61.83	76.33	67.97	73	90	83	29	82.00	44	55	50	2	51.67
Cleveland, Ohio	63.57	68.38	61.70	82	92	81	30	84.33	49	51	46	4	52.00
Columbus, Miss (6)	......	......	......	78	89	80	...	......	56	70	62	...	......
Concord, N. H	60.23	71.00	60.97	82	93	78	30	83.67	49	56	50	12	54.00
Cooper, Mich	61.72	69.93	58.90	86	90	75	29	82.67	50	52	46	4	51.33
Craftsburg, Vt	53.40	61.87	54.53	72	82	74	30	76.00	44	46	41	8	44.00
Crichton's Store, Va. (7)	71.55	81.88	80.24	82	99	94	30	91.67	56	62	67	4	63.34
Danville, Ky	71.19	79.85	69.31	85	92	79	29	83.00	58	61	58	5	60.00
Deaf & Dumb Inst. N. Y	66.23	72.92	66.05	81	90	85	29	84.54	55	57	57	19	57.60
Delaware College,(1)	64.21	75.84	66.50	78	92	76	22	75.67	52	60	58	5	60.67
Detroit, Mich	64.37	69.67	60.17	82	92	80	29	83.33	44	40	44	3	46.00
Dubuque, Iowa	63.41	75.66	65.07	79	91	80	29	83.33	45	52	51	3	53.00
Easton, Pa	63.60	77.90	65.80	78	*	83	30	86.67	56	58	53	19	58.00
Exeter, N. H	56.37	70.10	57.93	75	88	73	30	77.00	45	61	49	12	53.00
Flint, Mich	59.50	72·40	58.00	81	93	77	29	83.67	42	53	42	...	......
Fort Madison, Iowa.(8)	66.83	78.57	71.23	87	95	86	28	88.00	48	59	54	1	55.67
Frederick, Md	68.32	76.88	67.17	87	95	85	29	89.00	56	59	57	4	61.07
Friendship, Tenn	66.33	83.00	68.14	72	93	76	21	80.00	50	59	57	4	56.00
Gallipolis, Ohio	66.37	76.15	70.10	76	90	87	29	82.67	52	64	56	13	60.67
Gardiner, Me	63.70	71.68	60.37	78	91	72	29	79.67	54	57	51	20	56.33
Germantown, Ohio	64.17	74.30	63.05	78	91	79	29	84.49	46	53	47	2	49.10

(1) Incomplete.

*102

(2) Obs. taken at Sunrise, 2 P. M. and 10 P. M.

(3) For the last 16 days of the month only.

(4) For the last 22 days of the month only.

(5) Obs. taken at 6 A. M. 1 P. M. and 9 P. M.

(6) Mean for the month 75.41.

(7) Obs· taken at 7 A. M., Noon and 6 P. M.

(8) Obs. taken at 6 A. M. Noon and 7 P. M.

NAME OF STATION.	Mean.			Maxima.					Minima.				
	7 A. M.	2 P. M.	6 P. M.	7 A.M	2 P.M	9 P.M	Warmest day Date	Warmest day Mean temp.	7 A. M	2 P. M	9 P. M	Coldest day. Date	Coldest day. Mean temp.
Gettysburg, Pa	63.20	74.40	65.03	82	94	82	29	86.00	52	60	53	12	58.00
Glenwood, Tenn	64.25	76.86	68.02	75	89	80	29	79.97	49	60	55	4	55.57
Gouverneur, N. Y	59.90	69.53	59.30	76	88	76	30	80.00	42	50	48	12	47.33
Grand Rapids, Mich	56.07	70.07	59.40	78	89	78	29	81.33	40	46	41	2	42·00
Granville, Ohio	62.61	74.44	64.00	75	91	80	29	81.67	45	53	48	3	51.33
Great Falls, N. H	59.07	72.07	59.70	77	91	81	29	82.00	50	57	46	5	53.67
Green Springs, Ala	75.24	89.43	75.82	85	98	83	27	86.35	60	77	62	3	66.33
Hamilton, Ca.(1)	63.03	....	62.23	88	...	82	30	85.00	51	...	52	3	51·50
Harrisburg, Pa	66.73	77.11	70.79	85	97	90	30	89.33	57	62	60	4	62.00
Hillsboro, Ohio	61.43	72.20	62.72	75	88	77	29	79.83	45	50	46	3	52.33
Jackson, Ohio	....	....	....	...	...	...	...	....	...	...	...	...	....
(Crookham)	66.83	77.20	64.17	81	96	78	28	84.00	48	55	48	3	53.00
Jackson, Ohio	......	......	......	...	...	...	...	......	...	...	...	...	......
(Gilmor)	68.00	78.97	64.72	84	97	78	28	84.00	44	61	49	14	58.33
Jacksonville, Fa	76.85	82.75	75.50	82	89	82	21	82.67	68	75	68	4	72.67
Key West, Fa	79.07	86.77	....	81	93	...	8	86.50	77	79	...	4	78.00
Knox Hill, Fa	73.00	81.72	74.27	80	89	78	23	81.33	62	68	67	3	67.67
Knoxville, Tenn	66.22	77.47	68.25	78	89	79	28	80.33	54	61	55	3	58.67
Lac qui parle, Min	62.17	74.84	63.07	78	92	82	28	83.00	42	54	41	1	46.00
Lewisburg, Va	63.23	75.50	62.73	76	91	75	28	77.67	44	60	48	4	51.33
Lima, Pa	66.18	75.16	64.86	84	92	82	30	86.20	57	60	51	19	58.20
Lodi, N. Y	60·50	68.73	59.50	82	89	80	30	83.33	46	54	46	12	48.67
Londonderry, N H	58.43	72.83	60.20	85	94	82	30	87.00	47	54	48	7	51.67
Lowville, N. Y (2)	52.47	69.60	57.10	68	89	75	30	77.33	38	52	45	12	46.00
Madison, Ind	61.91	72.45	65.25	73	88	78	29	79.33	46	54	49	2	50.67
Madrid, N. Y	60.81	70.48	57.22	75	89	78	29	79.67	49	56	46	11	52.76
Manchester, Ill (3)	66.18	78.02	66.00	82	93	85	28	85.83	47	56	50	2	51.50
Manchester, N. H	61.33	74.00	64.06	80	96	83	30	86.33	49	59	53	12	56.67
Meadville, Pa	58.87	70.31	60.63	75	90	79	29	81.00	43	50	46	3	48.66
Mendon, Mass	62.50	70.87	62.10	81	92	80	30	84.33	53	58	52	8	56.00
Millersburg, Ky	65.73	74.50	68.70	79	88	81	29	82.67	48	56	48	3	54.67
Milton, Ind	62.85	74.05	63.57	79	90	79	29	82.67	47	47	44	2	46.17
Milwaukie, Wis	....	....	....	...	...	...	...	....	...	...	...	...	....
(Winkler)	57.30	66.90	54.93	78	90	78	29	81.00	40	48	37	2	42.67
Morrisville, Pa	63.03	75.07	64.03	76	92	79	30	81.33	54	56	55	12	60.00
Moss Grove, Pa	59.61	69.68	56.91	78	92	74	29	80.67	40	53	42	12	48.33
Muscatine, Iowa	63.34	75.63	64.03	82	90	84	16	84.33	35	53	47	2	50.33
Nantucket, Mass	63.38	68.32	59.53	76	84	70	30	76.00	57	62	53	8	59.00
Newark, N. J(4)	....	....	....	...	...	...	...	85.38	...	...	...	13	59.50
New Bedford, Mass	63.10	69.10	61.28	81	87	79	29	80.67	50	60	52	13	55.16
Newburyport, Mass	63.11	70.31	62.15	82	92	81	30	82.86	54	55	51	7	56.00
New Harmony, Ind	70.13	78.73	68.60	85	93	80	29	85.67	54	60	55	2	56.33
New London, Conn	64.27	73.40	62.37	81	93	80	30	84.00	54	62	54	13	59.33
New Orleans, La	73.57	84.10	77.84	76	89	80	28	80.83	65	72	71	3	72.50
New Wied, Texas	77.47	91.33	74.70	83	97	86	29	87.33	65	72	65	7	65.67
Norristown, Pa	62.79	73.74	66.13	80	92	82	30	83.60	54	58	54	19	57.03
North Attleboro,	....	....	....	...	...	...	...	....	...	...	...	...	....
Mass	62.61	71.41	60.55	80	92	79	30	81.90	49	59	51	13	56.26
Oberlin, Ohio	63.33	71.70	60.90	81	93	80	29	84.67	48	54	43	4	52.33
Oldtown, Me.(5)	51.10	63.14	51.74	60	74	57	2	62.33	41	52	46	9	49.67
Oswego, N. Y	55.60	64.07	58.47	72	92	78	30	80.67	36	44	40	7	47.00
Ottawa, Ill	66.67	75.13	66.63	83	94	90	29	88.33	47	58	44	2	50.67
Oxford, Miss	70.73	79.43	70.97	79	87	80	29	81.00	59	61	58	7	63.00
Pella, Iowa	60.50	79.17	62.23	72	96	76	27	79.67	48	54	49	1	51.33

(1) Obs. taken at 9 A. M. and 9 P. M.
(2) Obs. taken at Sunrise, Noon and 9 P. M.
(3) Obs. taken at 7 A. M, 1 P. M. and 9 P. M.
(4) Obs. taken at 7 A. M. and 6 P. M.
(5) For the first 19 days of the month only.

NAME OF STATION.	Mean.			Maxima.					Minima.				
	7 A. M.	2 P. M.	9 P. M.	7 A.M.	2 P.M	9 P.M.	Warmest day Date	Warmest day Mean temp.	7 A. M	2 P. M	9 P. M	Coldest day Date	Coldest day Mean temp.
Penn Yan, N. Y..(1)...	54.13	70.07	60.33	74	90	75	30	77.67	40	54	48	12	49.33
Perry, Me(2)...........	51.03	64.73	54.37	63	82	68	29	70.67	44	54	45	17	49.00
Perrysburg, Ohio(3)...	66.87	73.60	64.47	90	92	82	29	88.00	48	55	46	3	51.00
Philadelphia, Pa.......	67.70	76.60	71.30	85	94	88	30	88.67	58	60	60	19	60.00
Pittsburg, Penn.......	....	....	....	...	...	...	...	....	...	...	...	...	....
(Oakland Station.)...	61.50	68.67	60.63	80	91	78	30	82.67	44	50	43	12	47.67
Pittsburg, Penn.......	....	....	....	...	...	...	...	....	...	...	...	...	....
(Marine Hospital.) (4)	60.80	70.07	65.53	77	90	81	29	84.00	46	54	55	4	53.34
Platteville, Wis........	60.00	80.00	62.00	80	96	78	...	....	44	60	53	...	....
Pocopson, Pa............	65.77	76.84	66.45	83	96	83	30	87.00	55	62	56	4	60.67
Pomfret, Conn.........	60.79	69.34	61.43	80	88	79	30	81.33	48	58	52	12	53.67
Pottsville, Pa...........	61.28	73.73	63.22	83	96	78	30	85.90	41	56	52	4	53.83
Poultney, Iowa.........	63.17	77.57	60.53	78	94	77	28	82.67	43	53	40	2	46.33
Princeton, Mass........	57.90	66.90	58.30	78	85	75	30	78.20	45	52	46	5	49.33
Providence, R. I. (5)...	61.70	72.00	62.10	78	95	78	30	81.67	48	63	51	13	57.33
Quasqueton, Iowa.....	61.55	78.24	63.24	77	92	89	16	88.36	44	54	46	2	50.26
Randolph, Pa...........	59.63	70.63	58.07	82	93	79	29	84.33	44	53	43	12	48.00
Red River, ...	....	....	....	...	...	...	...	....	...	...	...	...	....
Settlement............	65.13	76.63	65.53	76	92	76	13	78.00	54	63	48	5	59.00
Sag Harbor, N. Y.....	63.20	75.57	62.67	78	94	83	30	85.00	54	62	55	13	60.33
St. James, Mich........	57.27	67.56	52.28	79	90	78	30	82.33	39	46	36	2	40.33
St. Johnsbury, Vt.....	58.83	68.73	58.40	78	91	78	30	82.33	46	53	45	13	52.00
St. Louis, Mo..........	67.68	78.50	68.78	81	95	82	28	85.83	52	60	53	2	55.50
St. Martins, C. E......	57.54	70.04	59.36	81	94	79	30	84.76	43	56	49	12	47.30
Saugatuck, Mich.......	63.26	70.33	60.43	82	92	81	29	83.33	46	58	42	2	47.00
Savannah, Ohio.........	62.50	72.00	60.47	84	97	78	29	86.00	43	50	44	3	47.00
Savannah, Ga..........	73.30	84.40	74.60	79	93	80	24	82.46	63	71	67	4	70.06
Saybrook, Conn........	60.97	68.17	62.17	71	83	78	29	75.33	51	58	56	12	56.33
Schellman Hall, Md....	65.50	75.23	65.60	83	91	89	29	87.67	55	59	55	18	56.67
Smithsonian Inst......	66.43	77.42	69.78	81	95	86	29	86.67	55	64	58	4	59.00
Smithville, N. Y........	54.60	64.70	55.87	69	84	76	27	72.33	43	52	45	12	47.00
Smithfield, Va. (6).....	....	....	....	...	...	...	29	85.75	...	...	...	13	60.50
South Thomaston,.....	....	....	....	...	...	...	...	....	...	...	...	...	....
Me......................	70.84	72.23	71.24	85	*	94	30	93.00	66	67	65	10	66.13
Sparta, Ga................	65.83	85.47	71.40	73	92	80	30	81.67	50	71	57	3	60.67
Spencertown, N. Y....	60.50	69.33	62.23	70	95	82	30	82.00	49	55	50	12	51.33
Springdale, Ky.........	58.35	80.12	69.20	69	96	86	17	81.67	40	55	50	2	53.50
Springfield, Mass......	62.26	73.54	64.13	81	92	86	30	86.33	49	58	52	12	55.33
Steuben, Me............	57.83	63.83	58.83	73	87	71	29	77.00	50	53	47	8	51.67
Superior, Wis...........	59.74	64.11	59.63	75	87	74	29	75.33	40	46	46	11	47.33
The Rock, Ga. (7)......	....	....	....	...	...	...	...	....	...	...	...	...	....
Unionville, Ohio........	63.13	68.90	59.57	81	92	80	30	82.33	52	54	46	3	51.33
Upper Alton, Ill........	67.20	76.79	67.10	78	92	82	27	82.67	50	59	52	2	55.33
Urbana, Ohio............	62.86	75.37	62.60	78	94	79	29	83.67	44	58	44	2	48.33
Wampsville, N. Y.....	61.00	73.52	59.72	78	94	83	29	84.00	49	54	45	12	52.33
Warrington, Fa..(8)...	75.93	81.00	80.17	82	84	82	24	82.67	66	72	74	8	73.33
Westfield, Mass........	56.79	70.70	61.47	77	93	81	30	83.33	45	56	48	13	52.00
W. Haverford, Pa.....	62.67	76.97	....	80	94	...	30	87.00	52	66	...	19	57.00
Whitemarsh Is. Ga.....	74.20	81.87	74.50	80	88	80	24	81.67	67	74	66	4	69.00
Williamstown, Mass...	55.44	69.02	61.99	79	87	78	30	81.50	44	55	48	9	53.33
Winchester, Va.........	67.17	77.87	66.33	80	94	84	29	86.00	56	56	52	4	59.33
Worcester, Mass.......	61.67	71.03	62.50	84	91	82	30	85.33	49	55	52	12	55.00
Zanesville, Ohio........	60.60	74.67	63.37	77	98	78	29	82.67	45	50	47	3	48.00

(1) Obs. taken at Sunrise, 2 P. M. and Sunset. (2) Obs. taken at Sunrise, 2 P. M. and 9 P. M.

(3) Obs. taken at 7 A. M. 1 P. M. and 9 P. M.

(4) Obs. taken at Sunrise, 9 A. M. 3 P. M. and 9 P. M. —Mean at 9 A. M. 64.97, Max. 81, Min. 50.

(5) Obs. taken at 6 A. M. 1 P. M. and 10 P. M.

(6) Mean for the month 73.74, Max. 93.5, Min. 50.5. *101

(7) Mean for the month 71.87, Max. 78 Min. 61.—Obs. taken at morning, 3 P. M. and evening

(8) Obs. taken at Sunrise, Noon and Sunset.

NAME OF STATION.	Mean. 7 A. M.	2 P. M.	9 P. M.	Maxima. 7 A.M	2 P.M.	9 P.M	Warmest day Date	Mean temp.	Minima. 7 A. M	2 P. M	9 P.M	Coldest day Date	Mean temp.
Alexandria, Va.	73.98	83.92	75.55	81	95	87	19	86.33	62	67	62	7	64.33
All Saints, S. C.	78.68	82.84	79.19	82	89	83	27	83.33	73	78	73	8	76.00
Amherst, Mass	67.39	78.08	68.20	79	92	77	1	81.67	57	65	61	22	58.33
Angelica, N. Y.	66.19	78.97	65.58	76	92	74	19	81.00	48	63	50	8	56.00
Ann Arbor, Mich. (Woodruff)	64.66	78.78	69.13	80	91	79	16	82.00	57	65	60	7	64.00
Ann Arbor, Mich. (Winchell)	68.25	77.03	66.44	81	88	77	16	81.40	59	65	57	20	61.60
Arcola, Ohio	70.68	76.61	69.22	82	90	78	19	82.00	60	66	56	21	62.00
Ashland, Va	71.42	84.32	72.81	79	93	80	19	83.67	58	66	64	8	65.33
Athens, Ill	75.41	86.09	75.41	85	101	84	18	89.33	65	74	67	21	70.00
Augusta, Ill	69.90	83.58	73.38	80	95	82	16	83.67	62	66	65	21	67.67
Austin, Texas	72.74	88·77	78.83	77	94	84	31	83.00	68	76	72	16	74.00
Baldwinsville, N. Y.	65.45	73.97	68.52	76	87	78	19	79·33	54	60	57	21	59·00
Battle Creek, Mich.	70.84	81.45	68.55	84	96	82	16	87.33	62	69	56	21	63.67
Bedford, Pa.	69.77	84.12	72.02	78	92	80	19	82.33	58	67	61	7	65.67
Beloit, Wis.	66.58	81,93	68.31	82	98	82	18	87.00	54	63	59	20	60.00
Beverly, N. Y.	69.90	77.13	71.20	77	88	79	1	80.67	61	63	60	21	61.67
Bladensburg, Md...(1)	68.33	86.23	72.70	80	102	78	19	86.17	64	70	62	8	67.00
Bloomfield, N. J.	71.60	82.47	71.56	81	96	80	19	86.50	59	62	60	21	63.66
Boston, Mass..(2)	67.87	79.89	71.13	86	90	84	19	86.67	56	68	60	22	63.67
Brandon, Vt.	67.01	74.10	68 00	80	89	79	1	82.34	53	59	56	8	59.83
Burlington, Vt.	69.03	79.87	68.49	80	92	80	1	82.33	59	66	57	8	64.67
Burlington, N. J.	72.81	82.65	74.71	82	94	86	19	87.33	60	66	62	22	63.67
Camden, S. C.	77.84	88.10	78.71	82	94	84	29	85.33	73	76	71	5	75·00
Canton, N. Y......(1)	68.04	79.04	70.26	76	91	80	1	80.33	57	64	60	7	62.00
Carmel, Me.	65.33	83.25	61.73	75	96	75	2	80.00	50	65	46	8	59.00
Cedar Keys, Fa.	78.71	83.90	79.00	82	91	82	16	83.33	74	76	74	4	76.33
Chapel Hill, N. C.	74.77	89.84	76.00	82	99	84	19	86.67	68	81	67	7	71.33
Charleston, S. C.	77.38	87.58	80.35	81	92	83	30	....	73	76	73	5	....
Chestertown, Md.	75.91	83.82	73.90	82	93	89	18	87.33	61	68	61	22	64.67
Cincinnati, Ohio..(3)	71.52	86.52	76.74	71	91	80	19	85.67	37	51	46	6	69.00
Cleveland, Ohio	70.90	78.03	70.29	83	91	80	19	84.67	60	66	59	21	64.67
Columbus, Miss..(4)	....	....	....	80	92	84	...	....	73	77	73	...	....
Concord, N. H.	68.80	78.90	68.93	82	93	80	1	84.33	58	67	60	22	63.67
Cooper, Mich.	70.87	76.06	67.97	84	90	80	19	80.00	58	62	56	20	61.33
Craftsbury, Vt.	62.74	73.39	64.54	74	82	74	17	74.00	52	63	56	21	58.00
Crichton's Store, Va.(5)	78.67	90.15	84.85	84	99	92	18	90.33	72	68	73	7	71.33
Danville, Ky.	76 52	85·03	75.03	83	94	80	18	85.00	70	69	68	6	69.33
Deaf & Dumb Inst N. Y.	73.17	79.32	72.94	86	93	84	19	87.40	59	61	61	21	60·40
Detroit, Mich.	70.65	79.32	69.26	84	91	82	16	84.33	60	64	60	20	62.67
Dubuque, Iowa	66.73	81.27	71.69	79	95	88	17	86.33	58	64	60	20	60.00
Easton, Pa.	70.77	85.87	74.52	78	96	84	19	85.33	56	62	61	21	62.00
Exeter, N. H.	64.35	75.58	65.68	75	90	78	17	79.67	53	65	50	22	60.33
Flint, Mich.	68.80	81.10	67.30	83	92	80	...	....	57	61	55	...	....
Flushing, N. Y(6)	....	....	....	...	...	...	...	....	...	...	...	...	....
Ft. Madison, Iowa..(7)	75.10	87.03	75.65	89	102	87	18	91.33	62	66	64	20	64.00
Frederick, Md.	75.50	84.86	75.02	85	93	83	18	87.33	59	71	64	22	63.00
Friendship, Tenn.	71.48	89.16	73.87	78	94	78	17	82.67	63	79	68	5	71.00
Gallipolis, Ohio	74.68	83.71	76.65	82	94	83	18	84.33	64	75	70	8	70.33
Gardiner, Me.	69.90	78.77	69.29	82	93	76	19	80.33	57	67	59	8	61.33
Germantown, O.	71.96	82.22	71.44	81	93	81	19	84.58	64	72	61	6	67.32
Gettysburg, Pa.	70.89	82.45	82.26	79	94	81	19	84.33	58	63	61	22	61.33
Glenwood, Tenn.	71.96	83.44	75.06	77	90	81	17	82.33	64	70	67	6	68.03

(1) Incomplete.
(2) Obs. taken at sunrise, 2 P. M. and 10 P. M.
(3) Obs. taken at 6 A. M., 1 P. M. and 9 P. M.
(4) Mean for the month 80.82.
(5) Obs. taken at 7 A. M. noon and 9 P. M.
(6) Monthly mean 76.45, Max. 96, Min. 60.
(7) Obs. taken at 6 A. M. noon and 7 P. M.

NAME OF STATION.	Mean.			Maxima.					Minima.				
	7 M. A.	2 P. M.	9 P. M.	7 A.M	2 P.M.	9 P.M.	Warmest day Date	Warmest day Mean temp.	7 A. M	2 P. M	9 P. M	Coldest day Date	Coldest day Mean temp.
Gouverneur, N. Y.	70.52	79.97	67.26	82	90	80	1	83.33	56	64	50	21	60.00
Grand Rapids, Mich.	66.16	78.13	68.10	80	93	82	16	84.67	55	64	51	20	58.67
Granville, Ohio	70.43	80.71	70.36	76	91	81	18	87.67	58	63	56	9	66.67
Green Springs, Ala. (1)	78.90	91.89	80.46	84	98	87	11	86.00	74	81	78	6	78.33
Harrisburg, Pa.	74.74	84.07	78.35	84	95	89	19	89.00	59	66	64	22	64.00
Hillsborough, Ohio	69.58	80.98	71.13	76	90	79	18	81.00	59	67	64	6	65.00
Jackson, Ohio	......	......	......	...	...	...	...	....	...	...	...	...	....
(Crookham.)	75.13	83.97	71.03	84	94	78	20	82.00	62	68	60	8	64.33
Jackson, Ohio. (Gilmor).	75.16	86.81	73.58	86	96	84	18	86.00	64	68	62	9	70.00
Jacksonville, Fa.	80.19	85.10	78.68	84	91	85	12	84.67	74	77	74	5	75.67
Key West, Fa.	78.71	84.06	....	81	88	...	13	84.50	72	73	...	21	73.50
Knox Hill, Fa.	76.07	83.54	75.37	79	88	79	16	81.33	71	72	71	2	73.00
Lacquiparle, Minn.	65.56	76.50	68.54	80	91	80	16	81.33	58	63	62	1	61.33
Lewisburg, Va.	70.35	82.71	70.81	77	91	78	19	79.00	61	73	60	8	65.33
Lima, Pa.	74.09	82.35	72.54	82	93	80	19	84.00	63	63	57	22	62.33
Lodi, N. Y.	69.03	76.61	69.19	81	90	80	19	83.33	58	61	57	7	62.33
Lowville, N. Y.	60.58	78.35	65.37	74	90	75	1	75 75	44	59	54	7	54.25
Manchester, Ill. (2)	71.75	85.40	72.35	78	96	81	16	83.67	64	74	62	21	68.67
Manchester, N. H.	69.55	82·79	71.77	83	96	83	1	87.00	58	65	60	21	64.33
Mendon, Mass.	70.16	77.68	69.71	91	92	94	1	87.00	58	64	55	21	63·00
Milton, Ind.	71.77	82.76	72.71	83	94	82	18	85.33	65	71	63	21	68·50
Milwaukie, Wis.	......	......	......	...	...	...	...	....	...	...	...	...	....
(Winkler)	65.19	75.68	62.19	79	94	79	18	84.00	54	57	53	20	55.33
Morrisville, Pa.	70.00	81.48	71.00	78	94	79	19	83.67	57	64	60	7	64.67
Moss Grove, Pa.	68.50	79.31	65.71	80	90	76	18	81.33	56	64	54	21	60.67
Muscatine, Iowa	68.42	81.87	68.87	81	95	80	17	84.67	55	65	60	20	62.33
Nantucket, Mass.	69.40	75.18	66.47	76	86	73	18	78.33	59	60	58	21	61.33
Newark, N. J. .. (3)	....	....	....	...	....	...	19	85.62	...	...	...	21	60.75
New Bedford, Mass.	69.08	75.07	67.74	78	89	75	19	80.17	61	64	59	21	62.67
Newburyport, Mass.	68.12	77.55	67.21	85	95	81	17	82.90	59	64	59	21	62·20
New Harmony, Ind.	77.06	85.42	76.42	85	95	83	19	87.67	69	75	69	6	71.67
New London, Ct.	69.71	78.48	69.19	79	89	79	19	81.00	60	66	62	21	63.33
New Orleans, La.	75.97	82.97	78.31	79	90	84	12	83.67	74	77	74	27	76.00
New Wied, Texas	78·45	94.52	77.94	82	99	85	7	87.67	74	88	72	16	77.33
Norristown, Pa.	72.80	81.53	72.86	87	91	80	19	83.67	60	60	61	21	62.00
North Attleboro, Mass.	69.19	79.97	66.74	85	93	80	19	84.33	58	65	56	21	60.00
Oberlin, Ohio	71.90	82.20	68.80	82	94	79	19	82.67	60	71	58	21	63.67
Oldtown, Me... (4)	63.00	77.00	65.20	72	84	78	18	75.67	51	70	57	22	60.00
Oswego, N. Y.	65.84	75.10	68.55	81	91	88	19	81.33	58	60	56	7	58.33
Ottawa, Ill... (5)	71.38	82.41	70.38	82	95	82	18	86.33	58	68	60	20	62.67
Oxford, Miss.	74.84	86.06	76.81	84	91	83	17	84.00	66	76	69	6	71-00
Pella, Iowa	67.06	84.74	69.22	88	98	82	17	89.00	59	70	60	20	63.33
Penn Yan, N. Y. (6)	61.74	76.90	68.90	74	90	80	19	81.33	48	59	58	21	59.00
Perry, Me.	56.68	68.84	64.10	66	78	68	19	71.33	50	56	51	9	59.00
Perrysburg, Ohio (2)	76.10	81.19	73.03	92	94	84	17	87.33	62	70	60	21	65.33
Philadelphia, Pa.	76.60	83.70	78.90	84	95	87	19	88.00	60	66	63	22	64.17
Pittsburg, Pa.	....	....	....	...	...	...	...	....	...	...	...	...	....
(Oakland Station) (7)	69.97	79.39	70.74	76	91	78	19	80.00	63	69	60	7	66.00
Pittsburg, Pa.	....	....	....	...	...	...	...	....	...	...	...	...	....
(Marine Hospital)	69.26	78.58	75.65	76	90	84	19	82.00	63	69	60	8	65.75
Plattsville, Wis.	69.80	84.97	72.93	85	96	86	18	88.67	56	66	61	1	63.67
Pocopson, Pa.	74.45	85.52	74.47	82	98	83	19	86.00	61	66	63	22	64·33
Pomfret, Conn.	67.73	75.87	67.41	80	88	78	19	82·00	57	63	58	6	69.67
Pottsville, Pa.	73.23	84.15	69.15	81	98	81	17	88.00	52	62	58	21	60.90

(1) Obs. omitted from the 14th. to the 27th. of the month.
(2) Obs. taken at 7 A. M. 1 P. M. and 9 P. M.
(3) Obs. taken at 7 A. M. and 6 P. M. Mean for the month 74.75. Max. 94.25. Min. 54.75.
(4) Incomplete.
(5) Obs. not always taken at the usual hours.
(6) Obs. taken at sunrise, 2 P. M. and sunset.
(7) Obs. taken at sunrise, 9 A. M. 3 P. M. and 9 P. M. Mean at 9 A. M. 72.94. Max. 83. Min. 63.

NAME OF STATION.	Mean.			Maxima.					Minima.				
	7 A. M.	2 P. M.	9 P. M.	7 A.M	2 P.M	9 P.M	Warmest day Date	Warmest day Mean temp.	7 A. M	2 P. M	9 P. M	Coldest day Date	Coldest day Mean temp.
Poultney, Iowa	68.58	83.69	65.61	82	100	80	16	85.67	55	63	53	1	58.33
Princeton, Mass.	65.76	74.50	69.02	77	91	75	1	79.00	55	60	55	21	57.33
Providence, R. I.(1)	69.04	80.04	69.00	78	95	81	19	84.33	60	68	61	21	63.33
Quasqueton, Iowa	67.18	83.91	68.91	80	96	82	16	85.00	56	64	59	20	62.00
Red River Settlement	66.68	73.00	67.58	87	92	82	24	78.33	56	66	50	7	62.00
Richmond, Mass	64.29	80.94	70·45	78	94	87	17	84.00	52	64	58	21	58.67
Sag Harbor, N. Y.	70.58	82.00	69.94	80	96	82	19	85.00	58	64	61	9	66.33
St. James, Mich	67.88	72.80	62.37	82	89	72	16	78.67	58	58	50	3	56.00
St. Johnsbury, Vt.	69.00	76.66	67.13	80	88	75	2	73.67	57	67	54	8	62.00
St. Louis, Mo	74.76	87 08	77.10	82	96	83	16	86.67	69	79	70	1	74.00
St. Martins, Ca. E.	67.51	81.29	67.57	77	92	78	1	82.00	52	64	54	7	59.80
Saugatuck, Mich	71.00	81.13	69.64	83	93	85	16	87.00	63	68	59	4	65.00
Savannah, Ohio	72.26	83.42	69.39	84	95	78	16	85.67	57	69	57	21	64.33
Savannah, Ga.	77.30	89.50	79.40	81	94	82	13	84.33	73	79	74	5	76.00
Saybrook, Conn	69.35	79.16	69.68	76	88	78	19	78.67	62	67	62	21	62.33
Schellman Hall, Md	73.45	81.94	72.74	82	91	85	20	85.67	59	66	60	21	64.67
Smithsonian Inst	74.47	85.47	76.98	81	95	85	19	86.33	65	71	67	7	67.67
Smithville, N. Y.	65.58	77.68	66.29	77	94	77	17	79.33	55	66	55	7	59.67
Smithfield, Va.(2)	....	....	....	...	...	...	19	84.25	...	...	...	8	69.25
Sparta, Ga	71.45	88.58	23.84	74	96	82	22	83.00	68	71	70	3	70.67
Spencertown, N. Y.	67.90	76.42	69.19	79	92	83	1	84.67	56	61	59	21	59.33
Springdale, Ky	67.18	87.76	76.94	77	97	83	17	83.33	55	70	65	6	68.00
Springfield, Mass	70.00	79.45	72.68	80	93	84	1	84.33	56	64	60	21	61.67
Steuben, Me	64.35	71.74	62.48	77	84	75	19	76.00	55	55	53	7	64.67
Superior, Wis	65.19	72.52	64.68	76	88	78	27	80.00	54	55	50	6	53.00
The Rock, Ga.(3)	....	....	....	...	...	...	...	....	...	...	...	...	....
Upper Alton, Ill	72.22	84.36	70.55	80	93	86	18	83.00	60	75	62	1	70.00
Urbana, Ohio	71.33	83.13	70.87	82	95	77	19	84.67	62	69	62	6	66.33
Wampsville, N. Y.	67.81	81.29	67.74	84	92	76	19	81.00	54	68	56	24	61.00
Warrington, Fa..(4)	77.45	82.74	82.00	82	88	84	17	83.67	72	76	79	27	77.00
Westfield, Mass	64.96	77.25	68.28	75	93	80	19	82.00	52	62	58	21	59.67
West Haverford, Pa	70.06	85.09	....	77	95	...	20	85.50	60	61	...	22	60.00
Whitemarsh Is, Ga	80.16	86.32	79.32	83	90	82	22	84.33	75	80	74	5	77.33
Williamstown, Mass(5)	65.86	76.49	67.81	77	86	76	25	74.00	54	64	58	21	60.00
Worcester, Mass	69.68	77.21	69.17	82	91	82	1	85.00	56	63	60	21	60.00
Zanesville, Ohio	69.19	84.35	71.42	78	95	80	19	82.67	58	64	58	9	65.33

(1) Obs. taken at 6 A. M. 1 P. M. and 10 P. M.
(2) Mean for the month 59.50. Max. 96, Min. 65.
(3) Mean foı the month 76.20, Max. 80, Min. 72. Obs. taken morning, 3 P. M and evening.
(4) Obs. taken at sunrise, noon and sunset.
(5) Obs. taken at 6½ A. M. 2 P. M. and 9 P. M.

NAME OF STATION.	Mean.			Maxima.					Minima.				
	7 A. M.	2 P. M.	9 P. M.	2 P.M	9 P.M	7 A.M	Warmest day Date	Warmest day Mean temp.	7 A.M	2 P.M	9 P.M	Coldest day Date	Coldest day Mean temp
Alexandria, Va	70.10	80.16	71.13	77	87	80	16	80.00	58	70	60	19	63.00
All Saints, S. C	77.61	83.53	79.06	81	92	85	11	85.33	68	74	71	20	67.67
Amherst, Mass	61.16	72.91	63.19	74	84	73	17	74.07	45	63	52	31	53.57
Angelica, N. Y...(1)	60.68	73.80	60.76	68	84	71	16	73.33	43	60	44	30	55.33
Ann Arbor, Mich	....	....	....	...	...	...	...	....	...	...	...	...	....
(Woodruff)	61.71	76.26	65.48	74	87	76	3	78.00	48	63	52	18	54.67
Ann Arbor, Mich	....	....	....	...	...	...	...	....	...	...	...	...	....
(Winchell)	63.57	75.21	62.63	77	84	78	3	78.30	50	62	46	18	52.67
Ashland, Va	68.35	80.93	70.19	76	90	78	4	80.33	54	67	56	18	59.33
Athens, Ill	71.71	82.93	72.06	88	98	82	3	87.33	56	69	60	17	62.00
Auburn, Ala	75.03	84.16	76.58	83	95	84	16	85.67	66	70	64	21	66.67
Augusta, Ill	65.26	78.06	69.03	80	92	82	3	85.33	50	67	57	17	58.33
Austin, Texas	73.81	88.19	80.39	78	96	88	5	87.33	70	76	71	25	73.33
Baldwinsville, N. Y	60.74	71.39	64.92	70	80	74	3	73.67	46	58	52	30	54.33
Battle Creek, Mich	65.48	79.22	66.71	77	92	80	3	82.67	52	66	53	18	59.00
Bedford, Pa	66.26	80.23	69.11	73	89	76	3	78.00	52	68	62	18	61.67
Beloit, Wis	61.90	78.03	66.45	76	89	79	3	81.33	44	64	54	18	56.67
Beverly, N. Y	65.16	73.41	66.32	75	80	75	16	75.67	51	64	56	31	58.00
Bladensburg, Md(1)	68.45	83.42	68.24	78	92	78	16	81.33	55	71	55	19	60.67
Bloomfield, N. J	65.16	77.53	65.76	77	86	77	9	78.33	51	65	55	31	57.00
Brandon, Vt	60.71	72.52	62.81	72	84	73	2	75.00	44	58	49	27	52.00
Burlington, Vt	61.58	74.94	63.84	72	90	74	1	77.00	46	59	52	27	53.33
Burlington, N. J	67.32	77.81	69.68	82	85	81	16	80.67	55	68	58	19	62.67
Camden, S. C	75.81	86.81	76.90	81	94	84	16	85.33	64	66	65	18	66.67
Canton, N. Y	61.74	76.19	63.77	74	93	78	2	80.67	47	60	46	30	52.00
Carmel, Me..(1)	56.20	75.33	55.00	70	89	77	3	73.00	42	63	40	31	49.00
Cedar Keys, Fa	79.97	84.84	81.39	84	91	85	9	85.00	74	75	74	20	74.23
Chapel Hill, N. C	71.77	84.10	74.16	78	98	81	13	83.67	61	64	62	18	62.67
Charleston, S. C	77.35	88.38	80.54	80	95	86	16	....	68	72	73	20	....
Chestertown, Md	69.90	78.77	69.26	80	86	77	16	79.67	58	70	58	19	62.67
Cleveland, Ohio	65.39	76.05	66.74	75	84	75	2	73.67	50	67	52	18	58.67
Columbus, Miss..(2)	....	....	....	80	93	84	2	....	70	73	67	...	....
Concord, N. H	62.63	71.66	64.43	76	85	75	4	75.00	46	64	51	31	54.67
Cooper, Mich	65.42	74.10	65.22	86	84	78	3	81.67	52	58	50	17	54.67
Craftsbury, Vt	55.52	65.85	59.16	66	82	70	3	72.00	40	54	44	31	49.00
Crichtons Store, Va.(3)	73.28	84.37	80.78	81	94	88	9	86.33	61	69	69	28	69.33
Detroit, Mich	66.70	76.00	66.60	77	88	77	3	80.6	53	65	52	18	57.00
Dubuque, Iowa	63.72	77.15	67.10	83	93	83	3	86.33	43	64	51	17	56.00
Deaf & Dumb Inst. N. Y.	69.54	75.54	68.82	76	84	78	16	78.60	57	64	60	28	61.56
Easton, Pa	64.74	80.48	65.39	79	90	79	4	81.67	50	62	54	8	61.00
Exeter, N. H	56.77	71.18	60.22	70	82	70	23	70.33	41	62	48	28	51.33
Flint, Mich	63.80	78.90	61.90	79	90	73	...	....	49	64	45	...	....
Flushing, N. Y(4)	....	....	....	...	...	...	...	....	...	...	...	...	....
Ft. Madison, Iowa.(5)	68.71	80.58	71.19	88	98	85	3	90.33	51	68	57	27	60.33
Frederick, Md	68.39	79.72	69.88	77	89	77	16	79.27	55	70	60	19	62.17
Gallipolis, Ohio	72.39	81.71	71.84	78	90	80	4	82.00	60	71	60	18	63.67
Germantown, Ohio	67.14	78.82	68.24	75	89	75	4	79.18	52	66	60	18	57.13
Gardiner, Me	61.45	73.31	62.14	71	86	72	4	75.00	50	63	52	31	56.33
Gettysburg, Pa	65.48	77.35	68.84	74	88	76	26	77.67	53	69	61	19	60.33
Glenwood, Tenn..(6)	72.58	81.67	74.82	79	88	80	13	82.00	65	74	70	27	71.80
Gouverneur, N. Y	62.42	77.35	64.39	80	90	80	3	81.33	46	62	48	19	58.00
Grand Rapids, Mich	61.52	75.52	65.87	71	90	78	2	77.00	46	57	48	18	53.67
Granville, Ohio	65.38	78.18	66.66	73	88	74	8	78.00	45	66	52	28	64.00
Green Springs, Ala	78.25	88.17	80.08	82	98	86	15	88.67	70	74	74	6	73.67

(1) Month Incomplete.
(2) Mean for the month 80.09.
(3) Obs. taken at 7 A. M. noon and 6 P. M.
(4) Mean for the month 70.44, Max. 86, Min. 51.
(5) Obs. taken at 6 A. M. noon and 7 P. M.
(6) Incomplete.

NAME OF STATION.	Mean.			Maxima.					Minima.				
	7 A. M.	2 P. M.	9 P. M.	7 A.M.	2 P.M	9 P.M.	Warmest day Date	Warmest day Mean temp.	7 A. M	2 P. M	9 P. M	Coldest day Date	Coldest day Mean temp.
Harrisburg, Pa	70.35	79.27	75 54	78	88	86	26	81.33	58	70	68	31	66.67
Hillsboro, Ohio	65.71	77.34	67.77	74	88	75	4	79.00	50	66	54	17	63.00
Jackson, Ohio	......	......	......	...	...	...	...	......	...	...	...	...	......
(Crookham)	68.84	81.94	66.48	77	90	74	16	78.33	52	69	50	18	57.00
Jackson, Ohio	......	......	......	...	...	...	...	......	...	...	...	...	......
(Gilmor)	70.73	84.10	73.73	80	95	83	31	83.33	50	70	55	18	58.33
Jacksonville, Fa	79.77	86.67	80.00	83	92	86	10	85.67	74	76	74	20	74.67
Key West, Fa	79.29	85.58	......	82	88	...	26	84.50	75	76	...	29	76.50
Knox Hill, Fa..(1)	77.60	84.60	76.27	84	94	84	15	86.00	72	75	72	19	72.33
Lacquiparle, Minn	58.84	70.00	61.48	72	81	73	19	71 67	46	56	49	16	50.33
Lewisburg, Va	69.13	80.20	69.48	76	86	76	23	78.00	52	68	54	20	60.67
Lima, Pa	67.40	77.10	67.02	77	84	76	16	77.20	55	66	54	19	60.70
Lodi, N. Y	62.58	73.22	63.06	72	81	72	3	74.33	50	59	45	30	52.67
Londonderry, N. H	63.48	75.68	64.68	75	85	75	22	74.67	44	64	50	31	51.67
Madrid, N. Y	66.85	72.77	62.92	79	85	74	3	78.00	46	55	44	27	50.83
Manchester, Ill...(2)	66.13	75.35	68.68	81	92	80	3	82.67	55	64	54	17	59.00
Manchester, N. H(1)	58.54	76.80	66.04	75	87	75	17	76.33	34	66	55	31	52.33
Meadville, Pa.....(1)	61.17	75.29	66.20	75	82	76	8	74.33	47	64	53	18	57.67
Mendon, Mass	62.03	75.00	63.84	76	84	78	16	75.67	46	66	50	30	55.33
Milton, Ind	67.13	80.80	68.63	79	91	83	3	83.00	50	67	52	18	56.83
Milwaukie, Wis	......	......	......	...	...	...	...	......	...	...	...	...	......
(Winkler)	63.13	73.35	59.74	72	90	71	3	79.33	49	50	47	26	53.00
Montreal, Ca. (3)	66.00	74.25	67.00	70	80	72	...	......	57	65	60	...	......
Morrisville, Pa.	65.00	77.74	66.35	75	86	74	16	76.67	54	64	58	31	62.00
Moss Grove, Pa	64.87	77.62	62.66	72	88	74	3	76.67	52	62	50	10	62.67
Muscatine, Iowa	66.97	78.90	65.48	77	96	77	3	83.00	53	61	52	17	58.35
Nantucket, Mass	67.70	73.21	65.01	73	80	72	16	75.50	59	62	56	31	60.67
Newark, N. J..(4)	......	......	......	...	...	...	17	77.38	...	...	...	19	61.00
New Bedford, Mass	63.52	72.13	63.36	72	82	74	17	75.80	49	61	52	31	55.17
Newburyport, Mass	61.36	74.40	63.52	73	86	73	23	76.70	49	61	53	28	56.40
New Harmony, Ind	71.52	82.68	73.18	82	90	81	4	83.00	59	70	61	18	64.33
New London, Ct	64.84	76.29	65.84	74	85	76	16	77.33	52	64	54	31	58.00
New Orleans, La	76.39	85.48	79.72	80	91	84	13	84.17	73	78	71	4	76.00
New Wied, Texas	79.03	94.33	78.00	83	102	84	7	88.67	72	81	73	24	75.67
Norristown, Pa	66.48	76.05	68.28	76	83	77	16	77.60	53	64	58	20	61.67
North Attleboro, Mass	61.29	72·44	59.22	71	83	75	4	71.57	44	59	44	31	51.40
Oberlin, Ohio	67.00	78.90	65.40	76	90	74	12	80.33	53	66	49	18	56.67
Oldtown, Me	57.50	71.92	58.19	68	78	65	24	70·33	45	65	46	31	53.67
Oswego, N. Y	62.39	74.64	65.45	75	84	75	16	75.67	48	65	52	30	58.67
Ottowa, Ill...(5)	66.03	79.35	66.83	79	93	82	3	84.00	55	65	55	26	62.67
Oxford, Miss. (6)	76.55	86.73	77.36	81	94	82	14	83.67	71	74	72	7	73.00
Pella, Iowa	61.58	80.13	65.35	71	98	80	3	83.33	46	61	48	17	54.33
Penn Yan, N. Y.(7)	56.35	73.77	66.33	69	84	74	3	73.67	44	60	54	30	56.00
Perry, Me. (8)	52.84	64.39	57·80	63	74	66	24	66.67	38	50	48	28	47.67
Perrysburg, (9)	69.51	79.90	70.20	80	92	80	4	83.00	58	68	56	18	60.67
Philadelphia, Pa	72.20	78.80	74.10	80	84	82	16	80.80	61	71	66	31	67.67
Pittsburg, Pa.	......	......	......	...	...	...	...	......	...	...	...	...	......
(Marine Hospital.)(10)	65.84	76.58	72.81	76	84	81	4	77.75	53	67	61	18	60.50
Pittsburg, Pa	......	......	......	...	...	...	...	......	...	...	...	...	......
(Oakland Station.)	65.84	76.74	67.81	74	84	75	4	76.67	51	65	55	18	58.67
Platteville, Wis	65.90	81.13	70.16	76	93	82	3	83.67	48	68	56	17	60.00
Pocopson, Pa	69.40	80.80	69.27	78	90	78	16	79.83	57	74	57	18	63.67
Pomfret, Conn	61.93	71.48	61.74	73	79	71	17	73.67	49	61	50	31	53.33
Port Gibson, Miss.(11)	77.67	86.67	78.75	80	92	89	29	83.67	74	71	70	26	72.67

(1) Incomplete.
(2) Obs. taken at 7 A. M. 1 P. M. and 9 P. M.
(3) For eight days only.
(4) Obs. taken at 7 A. M. and 6 P. M. Mean for the month 69.443., Max. 84.5. Min. 49.
(5) Hours of observation not perfectly uniform.
(6) For the first 14 days of the month only.
(7) Obs. taken at sunrise, 2 P. M. and sunset.
(8) Obs. taken at sunrise, 2 P. M. and 9 P. M.
(9) Obs. taken at 7 A. M. 1 P. M. and 9 P. M.
(10) Obs. taken at sunrise, 9 A. M. 3 P. M. and 9 P. M. Mean at 9 A. M. 69.90. Max. 77. Min. 59.
(11) For the last eleven days of month only.

NAME OF STATION.	Mean.			Maxima.					Minima.				
	7 A. M.	2 P. M.	9 P. M.	7 A.M	2 P.M	9 P.M	Warmest day Date	Warmest day Mean temp.	7 A.M	2 P.M	9 P.M	Coldest day Date	Coldest day Mean. temp.
Pottsville, Pa	66.06	78.79	64.80	74	91	74	13	77.17	53	64	53	19	60.17
Poultney, Iowa(1)	61.80	82.70	61.18	80	96	76	15	77.67	46	65	43	17	52.00
Princeton, Mass	59.84	70.27	60.53	71	79	70	16	72.50	45	60	50	31	54.17
Red River Settlement	58.90	69.26	61.65	70	86	77	5	76.33	44	56	44	15	51.33
Richmond, Mass	58.29	77.00	65.84	68	88	77	3	77.00	38	60	47	30	48.33
Sag Harbor, N. Y	65.64	80.29	65.40	75	89	75	9	77.67	55	64	53	31	60.00
St. James, Mich	63.00	70.82	62.53	74	84	78	3	76.00	53	57	52	27	51.67
St. Louis, Mo	70.02	80.72	72.34	80	93	82	3	85.00	60	70	63	17	64.00
St. Martins, C. E	57.84	76.38	60.80	74	93	74	3	79.00	36	57	40	30	48.00
Saugatuck, Mich	67.67	79.96	67.13	79	91	80	3	83.33	54	68	53	18	60.33
Savannah, Ohio	67.45	81.19	68.19	80	92	81	4	81.00	47	67	48	18	57.00
Savannah, Ga	77.10	88.70	79.50	80	97	85	11	87.17	68	72	71	20	70.43
Saybrook, Ct(2)	64.47	74.53	65.64	72	83	72	24	75.00	54	64	57	27	61.00
Schellman Hall, Md	68.16	77.93	68.74	76	88	79	16	80.00	55	68	58	28	62.67
Smithsonian Inst	70.56	81.22	72.98	80	92	81	26	82.00	58	69	63	19	65.67
Smithfield, Va.(3)	....	....	....	...	...	...	26	83.00	...	...	...	18	65.00
Smithville, N. Y	60.13	75.45	62.84	73	86	76	2	76.33	42	62	48	30	54.67
Sparta, Ga	71.35	87.53	74.77	75	95	80	16	83.33	64	69	66	20	67.33
Spencertown, N. Y	62.32	72.01	63.06	75	82	78	5	74.33	44	60	42	31	48.67
Springdale, Ky	64.68	85.32	76.11	73	94	81	12	81 00	45	71	63	18	61.33
Springfield, Mass	62.90	76.58	67.19	75	85	76	1	77.33	49	67	56	27	57.00
Stratford, N. H	50.53	70.87	60.06	65	86	73	16	72.33	33	56	48	30	47.33
Steuben, Me	60.77	69.03	59.48	67	79	70	24	70.00	47	60	45	31	52.67
Superior, Wis	60.32	69.06	60.13	76	87	70	6	75.67	50	50	52	15	53.00
The Rock, Ga.(4)	....	....	....	...	...	...	...	....	...	...	...	...	....
Upper Alton, Ill	67.29	78.59	68.26	78	91	81	3	82.00	52	65	60	17	61.33
Urbana, Ohio	68.81	82.56	68.61	78	95	76	22	82 00	51	69	51	18	57.67
Wampsville, N. Y	62.10	78.00	62.90	75	86	72	2	76.67	44	62	48	30	55.33
Warrington, Fa..(5)	77.84	83.13	82.61	83	88	87	27	85.00	71	75	77	5	76.33
West Haverford, Pa	65.03	78.96	....	74	88	...	16	80.00	52	64	...	31	64.50
Whitemarsh Island, Ga	79.55	85.84	79.39	83	92	85	11	86.00	68	72	72	20	70.67
Williamstown, Mass.(6)	58.99	72.47	62.09	73	82	75	16	76.23	40	62	46	31	51.83
Worcester, Mass	61.19	73.68	63.65	72	83	74	16	75.00	50	62	51	28	54.00
Zanesville, Ohio	64.84	81.10	69.42	73	92	75	2	79.00	52	70	56	19	60.67

(1) Incomplete.
(2) For the last 14 days of the month only.
(3) Mean for the month 76.92. Max. 92. Min. 60.
(4) Mean for the month 75.30. Max. 81. Min. 66. Observations incomplete, taken noon 3 P. M. and evening.
(5) Obs. taken at sunrise, noon and sunset.
(6) Obs. taken at 6½ A. M. 2 P. M. and 9 P. M.

NAME OF STATION.	Mean.			Maxima.					Minima.				
	7 A. M.	2 P. M.	9 P. M.	7 A.M	2 P.M.	9 P.M.	Warmest day. Date	Mean temp.	7 A.M	2 P.M	9 P.M	Coldest day. Date	Mean temp.
Alexandria, Va	63.97	76.00	67.31	78	90	82	11	82.00	46	56	51	28	55.00
All Saints, S. C.	74.10	82.55	77.62	81	90	84	18	84.33	64	73	72	28	73.00
Amherst, Mass	53.47	69.17	57.93	75	85	77	12	74.90	33	54	43	20	45.90
Angelica, N. Y ...(1)	57.56	68.14	57.45	68	84	72	12	75.67	40	52	40	19	46.33
Ann Arbor, Mich. (Woodruff)	59.20	70.83	62.30	72	87	74	1	77.67	40	57	45	27	49.67
Ann Arbor, Mich. (Winchell)	60.51	73.10	59.82	73	86	72	1	77.03	39	51	40	27	46.60
Ashland, Va.	66.13	77.27	68.00	75	87	76	17	78.67	46	64	50	27	55.33
Athens, Ill	68.10	79.30	69.53	78	91	78	1	81.67	50	63	52	27	58.00
Auburn, Ala	75.43	85.43	76.73	81	92	83	13	86.33	63	77	68	27	70.00
Augusta, Ill	64.33	76.43	67.36	74	91	78	1	80.00	39	59	50	27	53.33
Austin, Texas	71.40	85.80	77.20	76	92	82	2	82.33	54	76	68	30	70.00
Baldwinsville, N. Y.	55.13	65.50	59.73	72	82	75	12	76.67	37	48	43	19	44.00
Battle Creek, Mich.	62.33	73.77	63.39	78	89	77	1	81 00	47	60	47	27	52.67
Beloit, Wis.	58.40	72.06	59.80	71	90	72	8	76.33	40	52	46	27	48.67
Beverly, N. Y.	59.75	70.04	61.37	77	81	81	17	71.33	44	56	45	19	49.33
Bladensburg, Md...(2)	61.64	76.40	65.22	75	93	74	2	80.00	43	60	49	30	57.00
Bloomfield, N. J.	58.22	74.48	61.60	78	94	80	9	80.67	40	62	45	19	52.50
Boston, Mass......(3)	55.33	71.93	65.07	77	90	80	9	78.67	38	56	50	29	50.00
Bowling Green, Ky.	69.27	84.57	69.03	76	94	77	20	79.67	44	70	51	27	55.67
Brandon, Vt.	53.84	66.41	55.60	70	84	74	12	75.67	36	47	38	19	42.17
Burlington, Vt.	54.90	68.70	58.26	72	85	78	12	77.67	34	48	39	19	43.33
Burlington, N. J.	61.50	74.20	65.60	78	90	83	2	82.00	46	64	50	19	53.33
Camden, S. C.	71.80	84.76	74.80	78	93	84	12	84.33	59	74	61	28	65.67
Canton, N. Y	52.72	68.72	57.50	74	85	72	12	78.33	33	48	36	19	41.00
Carmel, Me.	49.79	67.70	50.14	75	89	73	9	76.67	32	47	36	19	42.67
Cedar Keys, Fa.	77.70	84.53	79.63	82	89	83	23	84.00	72	78	74	25	76.33
Chapel Hill, N. C.	66.83	82.13	71.73	75	94	82	13	83.33	52	66	56	28	62.00
Charleston, S. C.	74.83	84.70	78.50	80	92	83	18	....	68	74	70	28	....
Cincinnati, Ohio..(4)	68.20	78.70	70.97	77	89	82	12	79.33	50	60	56	28	60.67
Cleveland, Ohio	61.59	73.25	64.74	77	86	79	1	79.00	42	58	46	27	52.00
Columbus, Miss..(5)	....	....	....	78	92	82	...	....	57	72	60	...	....
Concord, N. H.	53.47	69.50	58.97	73	85	82	9	74.33	34	55	38	20	44.33
Cooper, Mich.	62.33	69.73	62.79	76	85	76	10	77.00	48	56	48	19	52.00
Craftsbury, Vt.	49.63	63.47	55.60	66	79	74	12	72.00	32	43	34	19	37.00
Crichton's Store, Va. ....(6)	67.25	77.75	74.62	82	95	88	18	85.67	55	64	57	20	61.67
Danville, Ky.	71.68	82.48	72.74	78	90	79	12	80.67	60	71	62	18	64.67
Deaf & Dumb Inst. N. Y.	62.82	71.60	64.12	79	87	80	12	80.67	46	60	50	19	51.96
Detroit, Mich.	61.80	71.13	62.80	74	87	74	1	78.00	41	55	47	27	50.00
Delaware College....(2)	60.68	73.79	64.90	78	86	80	12	79.67	44	60	50	28	53.00
Dubuque, Iowa.	61.40	69.10	63.33	75	91	76	10	79.67	46	54	48	18	52·33
Easton, Pa.	56.63	78.60	63.47	76	94	78	12	79.33	40	63	50	28	55.00
Exeter, N. H.	49.87	68.33	55.00	72	87	74	12	74.33	32	53	40	19	44.33
Flint, Mich.	59.00	73.70	59.30	72	87	72	...	....	39	55	39	27	....
Flushing, N. Y. (7)	....	....	....	...	...	...	...	....	...	...	...	...	....
Ft. Madison, Iowa...(8)	63.93	78.77	68.47	76	92	80	9	81.00	34	60	51	30	50.33
Frederick, Md.	62.04	74.76	66.47	76	89	79	12	79.83	45	58	50	28	53.67
Gallipolis, Ohio	69.00	80.01	71.27	79	99	88	17	81.33	49	66	51	27	57.33
Gettysburg, Pa.	67.00	72.73	64.10	76	88	77	2	78.00	44	55	48	19	51.67
Glenwood, Tenn.	69.86	79.92	71.60	75	88	78	12	79.23	51	67	53	27	57.70

(1) For the first 23 days of the month only.
(2) Record incomplete.
(3) Obs. taken at sunrise, 2 P. M. and 10 P. M.
(4) Obs. taken at 6A. M., 1 P. M. and 9 P. M.
(5) Meanfor the month 78.06
(6) Obs. taken at 7½ A. M., noon and 6 P. M.
(7) Mean for the month 63.73, Max. 88, Min. 44.
(8) Obts. taken at 6 A. M., noon and 7 P. M.

NAME OF STATION.	Mean.			Maxima.					Minima.				
	7 A. M.	2 P. M.	9 P. M.	7 A.M.	2 P.M.	9 P.M.	Warmest day Date	Warmest day Mean temp.	7 A.M.	2 P.M.	9 P.M.	Coldest day. Date	Coldest day. Mean. temp.
Gouverneur, N. Y......	55.90	69.53	61.23	74	88	72	12	77.33	28	49	40	19	40.33
Grand Rapids, Mich...	58.93	71.47	61.67	74	83	75	10	76.33	43	57	44	27	49.33
Granville, Ohio.........	65.71	76.21	66.00	75	85	74	10	76.67	45	64	46	28	54.67
Harrisburg, Pa.........	64.07	64.50	69.93	79	89	84	12	82.67	48	60	57	20	56.67
Hillsborough, Ohio....	63.97	73.40	65.83	72	83	73	1	75.00	43	55	48	27	52.33
Hiram, Ohio...........	59.57	68.27	58.57	72	81	73	11	75.00	42	54	40	19	46.67
Jacksonville, Fa.......	78.33	84.93	78.43	83	91	83	15	85.33	74	76	73	25	75.66
Jackson, Ohio.........	....	....	....	...	...	...	...	... ...	...	...	...	...	....
(Gilmor)	69.93	79.50	72.23	89	93	84	11	85.33	47	53	48	28	49.33
Key West, Fa..........	78.83	84.77	....	81	89	...	22	84.00	77	82	...	3	79.50
Knox Hill, Fa..........	75.30	86.40	77.27	80	93	82	23	83.67	70	79	72	27	74.33
Lacquiparle, Minn.....	55.42	68.59	57.15	73	88	76	7	79.00	40	42	42	18	44.00
Lewisburg, Va	62.03	76·97	63.73	74	90	72	13	76.33	42	64	50	28	50.67
Lima, Pa..................	59.52	72.72	62.62	76	85	78	1	74.50	41	60	47	24	50.50
Lodi, N. Y...............	58.13	68.30	58.10	75	84	75	12	78.00	35	49	37	19	42.33
Londonderry, N. H....	53.73	69.87	58.83	77	90	78	12	80.33	32	54	42	19	46.67
Manchester, Ill.(1).....	64.60	76.47	66.93	75	92	80	8	79.67	44	62	46	27	54.67
Manchester, N. H......	51.89	69.64	58.82	74	88	77	12	76.67	28	52	40	19	45.33
Meadville, Pa...........	60.67	70.10	63.37	73	85	76	1	76.33	42	60	49	28	51.33
Mendon, Mass..........	56.17	69.17	59.10	75	89	78	12	78.67	39	51	40	20	43.83
Milton, Ind...............	64.42	75.43	66.58	76	86	75	18	78.83	43	58	49	28	53.67
Milwaukie, Wis........	....	......	....	....	....	....	....	....	....	....	....	....	....
(Winkler)	58.60	68.07	57.67	73	89	72	8	76.33	42	53	42	27	46.67
Montreal, Ca.	56.52	66.58	60.72	74	83	78	9	74.50	44	51	43	29	49.00
Morrisville, Pa.	58.83	73.63	61·60	73	91	76	2	77.33	46	60	50	19	52.67
Moss Grove, Pa........	59.63	72.96	60.04	74	88	74	10	78.67	34	62	44	28	47.00
Muscatine, Iowa.......	64.60	73.77	64.60	77	92	84	8	83.00	35	50	50	30	48.00
Nantucket, Mass.	62.52	69.04	69.82	77	79	71	9	75.83	52	56	50	19	54.10
Newark, N. J.(2)	......·	....	....	...	...	...	13	78.87	...	...	...	19	51.12
New Bedford, Mass....	56.70	68.93	57.93	72	86	72	9	76.33	39	57	44	20	48.67
Newburyport, Mass.(3)	55.47	69.06	58.77	76	92	79	12	79.17	40	56	44	19	47.97
New Harmony, Ind....	70.10	80.20	72.17	77	89	80	19	81.00	51	65	54	27	57.67
New London, Ct.......	57.97	73.00	61.73	74	84	74	9	74·67	44	62	50	29	33.33
New Orleans, La	75.43	82.70	75.33	78	89	80	14	82.33	68	75	73	27	73.83
New Wied, Texas.....	76.67	90.47	75.80	81	97	81	4	85.67	68	78	72	27	73.67
Norristown, Pa........	58.38	71.23	62.35	75	86	78	12	77.33	44	59	50	19	52.33
North Attleboro, Mass.	52.92	67.84	52.21	74	87	73	12	74.83	36	52	36	19	45.33
Oberlin, Ohio...........	64.50	75.50	63.90	77	89	74	11	77.67	42	59	43	27	52.67
Oswego, N. Y.	57.77	69.50	61.20	69	88	81	12	79.67	37	55	43	19	47.67
Ottawa, Ill...............	63.93	75.48	64.40	79	88	78	9	78.33	45	58	50	27	53.67
Oxford, Miss.(4).......	74.64	81.55	72.50	83	90	81	12	85.33	56	67	56	27	60.33
Pella, Iowa.	60.33	75.60	62.33	72	91	74	8	77.67	36	56	44	30	49.00
Penn Yan, N. Y.(5)...	54.33	69.13	61.63	71	84	78	12	78.00	32	51	41	19	44.33
Perry, Me................	47·93	60.57	51.87	64	76	68	9	66.00	34	45	40	19	40.67
Perrysburg, Ohio......	67.43	74.73	66.97	88	90	79	1	82.00	50	57	51	27	53.33
Philadelphia, Pa	65.75	75.67	69.20	79	90	84	12	83.67	50	65	58	20	58.83
Pittsburg, Pa...........	....	....	....	...	...	....	...	....	...	....	...	...	....
(Oakland Station,)..	62.77	72.10	65.17	75	86	76	2	76.67	44	57	51	28	51.67
Pittsburg, Pa.	....	....	....	....	....	....	...	....	...	....	...	....	....
(Marine Hospital)(6)	64.47	73.10	69.41	77	86	80	12	78.00	45	58	54	28	52.25
Platteville, Wis........	62.37	74.87	64.73	76	92	76	10	79.67	42	60	48	27	50.67
Pocopson, Pa............	62.33	76.02	64.35	76	90	79	12	79.67	48	63	49	28	55.17
Pomfret, Conn..........	55.73	67.33	57.00	74	85	74	9	74,67	42	52	41	19	47.67

(1) Obs. taken at 7 A. M. 1 P. M. and 9 P. M.

(2) Obs. taken at 7 A. M. and 6 P. M. Mean. for the month 64.496 ; Max. 88.75, Min. 40.

(3) Obs. taken at 7 A. M. 1½ P. M. and 9 P. M.

(4) For the last 22days of the month only.

(5) Obs. taken at sunrise 2 P. M. and sunset.

(6) Obs. taken at sunrise, 9 A. M., 3 P. M. & 9 P. M., Mean. at 9 A. M. 67°, Max. 78°, Minimum 50°.

NAME OF STATION.	Mean.			Maxima.					Minima.				
	7 A. M.	2 P. M.	9 P. M.	7 A.M.	2 P.M.	9 P.M.	Warmest day Date	Warmest day Mean temp.	7 A.M	2 P.M.	9 P. M	Coldest day Date	Coldest day Mean. temp.
Pottsville, Pa.	59.45	72.40	60.03	78	93	71	9	77.67	40	58	44	19	49.83
Poultney, Iowa	58.96	73.12	58.79	75	97	75	11	79.00	33	51	40	27	48.67
Princeton, Mass.	54.38	65.65	56.08	73	81	74	12	73.80	41	48	39	19	42.80
Providence, R. I.(1)	56.30	72.40	57.10	78	88	76	10	77.00	38	56	41	19	44.33
Red River Settlement	53.77	68.10	56.67	70	82	70	5	73.67	40	56	40	30	48.00
Richmond, Mass.	54.63	72.07	61.00	70	89	76	12	78.33	28	50	37	20	41.00
Sag Harbor, N. Y.	60.85	75.85	58.85	77	90	75	9	78.67	45	58	50	20.	54.67
St. James, Mich.	59.87	65.47	57.38	73	85	72	11	76.67	46	51	46	27	50.00
St. Louis, Mo.	68.47	79.77	70.93	75	90	79	9	80.81	49	64	56	27	57.83
St. Martin, C. E.	51.43	68.78	55.33	70	80	76	12	77.80	32	52	36	19	42.00
Saugatuck, Mich.	64.83	75.93	64.93	78	87	80	11	80.67	48	63	47	27	53.67
Savannah, Ohio	68.83	76.50	66.00	77	90	80	2	81.00	40	63	48	28	53.00
Savannah, Ga.	73.60	85.10	77.30	78	96	84	17	85.00	67	75	70	28	71.00
Saybrook, Ct.	59.70	70.00	59.27	72	80	74	9	74.33	44	60	42	20	52.33
Schellman Hall, Md.	63.07	75.40	65.10	78	89	79	12	81.00	45	60	50	30	55.33
Smithsonian Inst.	64.20	77.33	68.70	77	93	80	12	83.33	47	57	55	19	56.67
Smithfield, Va.(2)	....	....	....	...	...	...	13	83.50	...	...	...	28	57.50
Smithville, N. Y.	53.60	70.73	57.13	73	85	76	12	78.00	34	57	38	28	43.67
Sparta, Ga.	68.50	85.33	74.00	72	94	78	15	80.67	59	79	65	28	69.67
Spencertown, N. Y.	53.97	68.40	58.00	75	84	77	12	77.33	30	50	39	19	43.33
Springdale, Ky.	65.12	81.63	73.70	77	90	81	1	79.00	42	68	61	27	58.30
Springfield, Mass.	56·30	72.33	62.00	76	84	80	12	79.00	39	57	43	19	48.67
Stratford, N. H.	47.40	65.07	57.13	68	80	75	12	71.67	30	46	42	19	41.67
Steuben, Me.	53.90	62.17	52.70	69	84	71	12	68.33	37	49	42	23	44.33
The Rock, Ga. (3)	....	....	....	...	...	...	...	....	...	...	...	...	....
Unionville, Ohio	63.43	70.63	62.73	76	84	76	7	78.00	50	56	44	19	52.00
Upper Alton, Ill.	66.14	77·45	67.57	77	86	74	16	77.33	44	62	51	27	54.00
Urbana, Ohio	63.60	77.41	66.24	76	90	75	9	78.67	46	55	48	27	53.67
Wampsville, N. Y.	56.67	71.40	57.57	74	85	78	12	78.33	38	56	39	19	46.33
Warrington, Fa.(4)	77.30	83.13	82.03	81	87	86	20	83.67	72	77	76	27	75.33
West Haverford, Pa.	60.97	74.97	....	76	91	...	13	82.50	48	64	...	21	56.50
Whitemarsh Island, Ga	75.53	82.37	76.90	81	90	81	17	83.33	69	75	69	28	71.00
Williamstown, Mass.	53.35	66.65	57.27	72	79	72	9	71.33	33	50	40	19	44.40
Worcester, Mass.	55.97	67.07	56.77	78	86	75	12	76.67	34	42	40	19	42.00
Zanesville, Ohio	65.23	76.48	67.25	74	90	77	9	79.00	48	58	52	27	55.67

(1) Obs. taken at sunrise, 1 P. M. and 10 P. M.
(2) Mean. for the month 71.71, Max. 93, Min. 48. Obs taken morning, 3 P. M. and evening.
(3) Mean for the month 74.14 max. 78 Min. 70.
(4) Obs. taken at sunrise noon and sunset.

NAME OF STATION.	Mean.			Maxima.					Minima.				
	7 A. M.	2 P. M.	9 P. M.	7 A.M	2 P.M	9 P.M	Warmest day Date	Warmest day Mean temp.	7 A.M	2 P.M	9 P.M	Coldest day Date	Coldest day Mean temp.
Alexandria, Va.	46.13	59.42	50.13	65	72	65	1	65.67	28	44	35	25	38.00
All Saints, S. C.	56.87	71.10	62.73	77	80	78	1	77.67	40	55	42	25	45.67
Amherst, Mass	45.21	55.20	48.41	64	73	67	6	65.33	30	42	35	25	38.67
Angelica, N. Y.	41.14	52.50	42.68	59	75	58	5	61.00	22	32	26	26	28.67
Ann Arbor, Mich.	......	......	......	...	...	...	...	......	...	...	...	...	......
(Woodruff)	39.74	53.03	44.55	52	71	62	10	58.33	27	36	29	25	32.00
Ann Arbor, Mich.	......	......	......	...	...	...	...	......	...	...	...	...	......
(Winchell)	40.24	53.01	42.81	52	71	60	10	59.73	26	35	28	25	30.60
Ashland, Va.	43.74	60.94	50.64	64	76	60	1	62.00	28	40	32	25	36.00
Athens, Ill.	44.87	62.19	49.58	61	81	67	10	68.33	25	41	30	24	35.00
Auburn, Ala.	55.10	71.19	60.42	65	81	70	10	69.33	34	45	36	24	41.33
Augusta, Ill.	41.61	68.58	49.03	58	80	65	10	67.00	21	38	31	24	33.00
Austin, Texas	54.61	75.87	60.58	70	84	76	4	76.00	34	56	40	24	44.67
Baldwinsville, N. Y.	44.10	51.71	46.77	58	66	63	5	61.00	32	34	34	25	34.33
Battle Creek, Mich.	42.35	54.61	45.61	54	70	63	10	63.67	26	36	32	24	32.00
Bedford, Pa. (1)	42.55	54.00	54.07	70	82	71	1	72.67	27	34	32	25	32.66
Beloit, Wis.	38.64	54.71	42.13	57	74	60	10	63.67	20	37	29	25	30.67
Beverly, N. Y.	47.35	56.55	49.58	65	72	66	5	65.67	34	39	36	25	38.67
Bloomfield, N. J.	47.93	57.58	49.45	67	71	67	1	67.83	30	42	35	29	43.00
Boston, Mass. (2)	46.52	62.34	50.58	62	76	70	6	66.67	32	32	36	28	41.33
Bowling Green, Ky.	44.45	70.06	50.52	60	89	63	30	64.33	27	36	33	25	32.33
Brandon, Vt.	43.98	52.47	45.97	60	70	62	2	63.83	28	34	34	25	35.00
Burlington, Vt.	45.97	54.61	47.77	64	75	65	2	66.67	30	35	35	24	35.00
Burlington, N. J.	50.32	61.10	51.84	74	76	66	1	72.00	32	42	34	26	39.00
Camden, S. C.	50.68	71.90	55.61	72	83	69	1	74.67	31	57	34	26	43.00
Camden, Ark(3)	56.83	69.17	60.00	68	81	76	4	70.33	38	55	46	7	50.67
Carmel, Me. (1)	45.36	54.56	43.65	56	67	63	6	61.63	31	41	31	26	37.00
Canton, N. Y.	44.58	53.87	46.06	62	75	65	1	64.67	31	34	32	25	33.00
Castleton, Vt. (4)	39.48	52.88	42.72	54	67	58	20	59.33	28	37	34	24	36.00
Cedar Keys, Fa.	64.58	73.22	69.52	78	82	82	1	80.67	42	56	50	25	49.33
Chapel Hill, N. C.	50.42	68.74	56.35	68	83	69	1	71.67	31	44	38	25	43.00
Charleston, S. C.	58.58	72.19	63.96	76	86	74	1	......	39	56	44	25	......
Chestertown, Md.	48.23	58.93	48.94	66	72	62	1	65.33	30	41	33	25	36.33
Cheviot, Ohio	44.39	61.48	55.26	61	78	69	20	66.33	28	37	34	25	34.00
Cincinnati, Ohio. (5)	48.10	63.35	53.10	63	76	63	10	63.33	32	37	41	25	48.33
Cleveland, Ohio	45.00	54.30	47.40	61	71	61	10	62.00	32	35	35	25	34.67
Columbus, Miss. (6)	......	......	......	63	78	71	...	......	31	49	38	...	......
Concord, N. H.	45.42	56 07	48.43	62	72	66	5	64.33	32	40	35	29	38.33
Cooper, Mich.	45.55	52.45	46.00	62	76	66	10	68.00	30	35	31	24	32.00
Craftsbury, Vt.	40.71	49.90	43.29	58	71	60	3	63.00	27	33	28	25	31.00
Crichton's Store, Va(7)	52.28	64.78	64.13	70	79	77	1	73.00	36	43	45	24	44.00
Deaf & Dumb Inst.	......	......	......	...	...	...	...	......	...	...	...	...	......
N. Y.	50.29	57.59	52.04	67	70	66	1	67.40	34	42	37	25	39.83
Detroit, Mich	41.97	53.94	45.87	56	71	60	10	62.33	28	36	31	24	33.00
Delaware College. (1)	46.96	58.78	49.20	65	73	65	1	67.33	31	45	41	25	39.67
Dubuque, Iowa	41.55	57.10	46.70	55	81	66	18	66.00	26	33	30	23	31.33
Easton, Pa.	42.39	61.42	46.61	59	80	62	1	65.67	33	42	35	24	38.00
Exeter, N. H.	55.19	54.94	46.42	60	70	63	6	61.33	30	37	31	29	34.33
Flint, Mich.	37.60	51.90	42.20	50	73	58	...	......	25	31	26	...	......
Flushing, N. Y. (8)	......	......	......	...	...	...	...	......	...	...	...	...	......
Fort Madison, Iowa, (9)	40.03	59.55	49.61	52	78	67	10	65.67	18	38	32	25	32.67
Frederick City, Md.	45.93	59.00	49.37	62	76	63	1	64.97	31	40	36	25	39.10
Gallipolis, Ohio	46.97	62.22	49.19	60	74	59	1	62.33	31	43	32	25	35.33
Germantown, Ohio.	41.29	58.34	45.36	56	75	60	1	59.45	27	37	30	25	32.29

(1) Record incomplete.
(2) Obs. taken at 6 A. M. 1 P. M. and 10 P. M.
(3) For the first 12 days of the month only.
(4) For the last 25 days of the month only.
(5) Obs. taken at 6 A. M., 1 P. M. and 9 P. M.
(6) Monthly mean 59.56.
(7) Obs. taken at 8 A. M., noon and 5 P. M.
(8) Mean for the month 52.85, Max. 71, Min. 32.
(9) Obs. taken at 6 A. M., noon and 7 P. M.

NAME OF STATION.	Mean.			Maxima.					Minima.				
	7 A. M.	2 P. M.	9 P. M.	7 A.M.	2 P.M.	9 P.M.	Warmest day Date	Warmest day Mean	7 A. M	2 P. M	9 P. M	Coldest day Date	Coldest day Mean. temp.
Gettysburg, Pa........	44.06	58.48	47.39	58	72	61	1	62.67	30	40	32	25	36.00
Glenwood, Tenn.	48.56	63.56	51.92	62	75	65	28	62.83	32	36	34	24	34.87
Gouverneur, N. Y.....	42.90	54.13	47.68	60	70	60	6	62.00	28	32	30	25	31.33
Grand Rapids, Mich....	38.03	52.48	44.00	50	78	62	18	61.33	19	34	28	24	28.00
Granville, Ohio........	43.36	59.29	47.93	65	71	62	1	62.67	29	41	34	24	37.00
Green Springs, Ala.(1)	51.44	73.06	60.29	65	84	72	11	70.67	32	48	39	25	41.00
Harrisburg, Pa........	49.10	58.39	51.69	65	73	66	6	63.00	34	43	40	25	40.67
Hillsborough, Ohio ...	43.26	57.53	47.94	58	71	60	10	60.00	27	34	31	24	33.00
Hiram, Ohio...........	42.77	52.71	43.06	52	70	60	1	62.00	28	32	30	25	29.67
Jackson, Miss.	49.61	68.58	58.74	64	76	69	17	67.67	28	48	36	25	37.67
Jackson, Ohio, ...	......	......	......	...	...	...	...	......	...	...	...	...	......
(Gilmor)............	41.32	65.33	49.70	65	78	62	1	67.00	40	50	35	7	43.00
Key West, Fa.........	75.06	80.35	......	81	89	...	7	84.50	68	67	...	27	71.00
Knox Hill, Fa.	58.40	74.70	64.00	71	83	73	19	72.67	46	53	42	25	44.00
Jacksonville, Fa.	62.16	72.11	64.12	77	86	77	1	80.00	45	58	44	25	49.00
Lewisburg, Va........	44.70	61.04	45.65	64	74	56	1	60.67	26	34	31	26	33.00
Lima, Pa.	46.14	58·06	48.48	66	71	61	1	66.10	27	43	37	26	38.50
Lodi, N. Y.............	42.00	52.52	43.79	60	70	62	5	63.67	58	84	62	25	30.00
Londonderry, N. H....	46.06	57.35	48.35	67	75	66	4	66.00	32	39	35	28	38.33
Manchester, Ill.........	43.52	61.55	47.81	58	84	62	10	66.50	24	32	32	24	31.33
Manchester, N. H...(2)	49.36	62.81	54.72	62	72	66	3	65.00	37	56	39	8	40.33
Meadville, Pa..........	42.48	54.29	44.74	59	74	64	5	60.00	29	36	31	25	33.00
Mendon, Mass.........	50.42	57.71	51.71	66	69	64	3	65.00	34	45	37	26	39.33
Millersburg, Ky.......	44.77	60.52	46.48	59	78	65	4	63.00	29	36	34	24	35.33
Milton, Ind............	58.74	46.97	49.74	56	73	63	17	65.00	29	38	32	24	35.33
Milwaukie, Wis.......	......	......	......	...	...	...	...	......	...	...	...	...	......
(Winkler)...........	38.42	53.03	42.39	55	78	66	10	63.00	22	35	26	24	28.33
Milwaukie, Wis.......	......	......	......	...	...	...	...	......	...	...	...	...	......
(Pomeroy)....	41.74	59.23	43.77	60	86	64	9	67.67	22	36	28	24	28.67
Montreal, Canada......	45.93	52.31	47.41	62	70	65	5	63.45	32	34	33	24	34.67
Morrisville, Pa.	47.00	55.84	49.03	66	74	64	1	67.33	32	43	38	25	39.67
Moss Grove, Pa........	41.66	52.29	42.39	58	69	62	1	58.33	24	34	29	25	30.67
Muscatine, Iowa.......	38.10	58.71	44.52	51	84	61	10	64.33	15	36	30	24	31.00
Nantucket, Mass......	55.84	61.03	55.07	64	69	66	3	64.73	43	49	43	29	45.67
Newark, N. J..(3)......	......	......	......	...	...	...	1	67.25	...	...	...	26	40.38
New Bedford, Mass....	51.26	59.17	51.88	63	70	66	6	65.70	36	48	38	29	42.70
Newburyport, Mass.(4)	47.16	56.22	49.01	59	75	66	5	65.40	33	39	37	28	38.60
New Harmony, Ind....	45.82	62.06	50.58	61	75	66	20	63.33	28	42	35	25	35.67
New London, Conn....	50.19	60.77	52.65	64	74	67	3	67.00	36	49	37	26	44.67
New Orleans, La.......	60.02	69.47	64.35	73	77	72	5	72.83	39	50	46	25	45.00
New Wied, Texas......	59.87	76.77	64.87	79	89	87	4	84.33	37	60	47	24	49.33
Norristown, Pa.........	46.08	60.61	49.98	66	71	61	1	65.33	31	40	37	25	39.00
North Attleboro,.......	......	......	......	...	...	...	...	......	...	...	...	...	......
Mass	44.42	57.60	45.50	60	72	60	3	62.10	30	46	32	17	36.66
Oberlin, Ohio..........	43.70	55.97	45.29	83	72	60	1	64.00	30	39	30	25	33.67
Oswego, N. Y..........	45.87	54.77	48.55	60	71	60	5	61.33	33	37	38	24	37.67
Ottawa, Ill...(5).........	40.07	59.87	46.71	53	82	64	10	66.00	22	38	34	23	36.00
Oxford, Miss...........	51.68	66.16	54.68	63	77	68	20	68.33	28	38	29	6	45.33
Pella, Iowa............	37.10	57·35	42.53	51	79	61	18	64.00	20	32	23	24	26.67
Penn Yan, N.Y...(6)...	42.23	53.97	47.03	58	75	68	5	64.67	31	34	33	25	33.67
Perry, Me..............	43.61	51.10	44.87	58	62	58	6	56.67	30	40	29	29	33.67
Perrysburg, Ohio......	45.90	59.06	51.00	63	77	66	10	68.33	29	40	36	13	42.00
Philadelphia, Pa	50.84	53.69	53.69	71	77	70	1	72.16	35	43	41	25	41.67

(1) Obs. taken at sunrise, 2 P. M. and 9 P. M.
(2) For the first 11 days of the month only.
(3) Mean for the month 51.43., Max. 72½°, Min. 33°, Obs. taken at 7 A. M. and 6 P. M.
(4) Obs. taken at 7 A. M., 1½ P. M. and 9 P. M.
(5) Record incomplete for the last eight days of the month.
(6) Obs taken at sunrise, 2 P. M. and sunset.

NAME OF STATION.	Mean.			Maxima.					Minima.				
	7 A. M.	2 P. M.	9 P. M.	7 A.M.	2 P.M.	9 P.M.	Warmest day Date	Warmest day Mean temp.	7 A.M	2 P.M	9 P. M	Coldest day Date	Coldest day Mean. temp.
Pittsburg, Pa. (Oak Station)	44.06	55.52	47.13.	60	70	60	1	61.00	30	38	32	25	35.33
Pittsburg, Pa. (Marine Hospital,)(1)	45.42	54.77	50.86	61	68	65	1	62.75	34	38	37	25	36.75
Platteville, Wis	39.87	59.03	45.90	56	77	65	...	64.33	20	34	20	...	25.67
Pocopson, Pa	47.85	59.36	49.83	67	75	63	1	67.67	33	43	38	25	40.33
Pomfret, Conn	28.90	45.94	55.70	64	71	65	6	64.33	32	44	34	26	38.33
Portland, Me	21.59	54.71	47.79	57	69	60	5	61.00	32	41	38	28	39.00
Pottsville, Pa	44.89	54.72	43.48	65	68	57	1	63.70	26	38	32	26	34.33
Poultney, Iowa	37.92	55.83	41.48	50	81	61	18	63.33	24	30	29	24	29.00
Princeton, Mass	44.74	53.36	46.35	61	70	63	2	62.00	29	38	33	28	36.17
Providence, R. I (2)	47.60	59.80	49.90	64	74	66	6	67.33	33	48	37	29	40.33
Quasqueton, Iowa (3)	36.28	57.31	43.24	47	79	58	10	57.00	16	32	28	23	30.00
Red River Settlement	38.14	49.72	38.93	63	70	60	10	62.33	19	28	20	20	26.00
Richmond, Mass	44.32	56.42	49.23	63	76	66	2	66.00	29	39	34	29	34.00
Sag Harbor, N. Y	50.96	63.68	53.00	66	78	67	1	71.66	38	49	41	25	44.00
St. James, Mich	41.32	51.31	41.71	61	68	60	1	63.00	31	32	32	28	33.00
St. Louis, Mo	45.89	63.16	52.74	56	80	65	10	67.00	28	35	35	24	34.00
St. Martins, Ca. E	41.43	53.26	44.62	58	69	63	6	62.43	28	33	32	24	31.93
Saugatuck, Mich	46.00	57.06	47.81	62	78	66	10	68.67	28	40	31	24	33.00
Savannah, Ohio	41.29	57.13	45.32	60	76	61	10	63.00	27	37	35	25	33.00
Savannah, Ga	56.50	73.40	62.50	74	86	74	1	77.36	37	55	46	25	46.20
Saybrook, Conn	48.10	57.65	53.45	65	68	68	6	67.00	35	46	38	25	42.00
Schellman Hall, Md	46.71	62.26	48.48	60	76	68	1	66.00	34	41	34	24	38.00
Smithsonian Inst	48.26	61.39	52.18	67	77	66	1	68.33	32	44	38	25	40.33
Smithfield, Va (4)	....	....	....	...	...	...	1	70.75	...	...	...	25	42.00
Smithville, N. Y	42.77	52.71	44.58	60	68	61	5	62.33	28	32	31	24	31.67
Sparta, Ga	48.06	71.00	56.55	70	81	70	1	70.67	29	48	33	25	38.00
Spencertown, N. Y	43.65	53.06	46.90	64	70	64	1	64.33	27	38	35	25	36.33
Springdale, Ky	40.74	64.00	57.10	62	78	68	30	67.83	25	42	37	24	35.83
Springfield, Mass	45.81	58.58	49.26	66	77	66	3	67.33	30	43	36	25	39.33
Stratford, N. H (5)	41.58	51.03	44.55	58	67	64	3	63.00	28	37	33	26	37.00
Steuben, Me	45.94	52.87	46.39	60	62	59	4	59.33	31	44	29	26	36.00
Superior, Wis	39.06	49.58	42.39	58	68	63	8	57.67	24	28	24	23	26.67
The Rock, Ga (6)	....	....	....	...	...	...	...	....	...	...	...	...	....
Unionville, Ohio	46.94	54.03	47.40	62	70	58	5	61.33	32	38	30	24	36.00
Upper Alton, Ill	43.61	59.07	50.43	58	75	64	31	59.33	26	34	34	24	33.00
Urbana, Ohio	42.03	59.42	44.87	62	75	60	1	59.67	29	34	31	24	32.67
Wampsville, N. Y	44.42	54.06	45.16	62	75	64	5	64.33	32	37	32	25	33.67
Warrington, Fa (7)	61.58	71.71	70.16	78	82	51	5	76.67	39	51	52	25	48.00
Westfield, Mass	43.68	55.26	47.52	64	74	66	5	64.67	30	41	35	27	39.00
W. Haverford, Pa	47.25	59.58	....	66	75	...	1	69.00	31	34	...	26	39.00
Whitemarsh Is, Ga	57.52	70.90	62.10	75	79	73	1	75.33	39	54	47	25	46.67
Williamstown, Mass	43.73	52.58	45.36	62	70	62	2	64.56	28	37	32	25	35.20
Winchester, Va	47.19	60.61	45.68	62	76	64	5	64.33	30	44	36	26	39.33
Woodward High School, Ohio	47.13	62.52	50.23	64	75	63	10	66.00	34	42	30	24	39.67
Worcester, Mass	46.03	56.16	48.42	62	72	65	6	65.00	32	41	35	29	37.67
Zanesville, Ohio	43.58	59.68	49.06	56	70	68	5	65.33	32	36	32	25	33.33

(1) Obs. taken at sunrise, 9 A. M., 3 P. M. & 9 P. M. Mean. at 9 A. M. 47.10, Max. 61°, Minimum 35°.
(2) Obs. taken at sunrise, 1 P. M. and 10 P. M.
(3) Record incomplete.
(4) Mean for the month 57.18, Max. 78, Min. 32.
(5) Obts. taken at morn., noon and night.
(6) Mean. for the month 58.85, Max. 72, Min. 37.—Obs taken morning, 3 P. and evening.
(7) Obs. taken at sunrise noon and sunset.

NAME OF STATION.	Mean.			Maxima.					Minima.				
	7 A. M.	2 P. M.	9 P. M.	7 A.M	2 P.M	9 P.M	Warmest day Date	Warmest day Mean temp.	7 A. M	2 P. M	9 P. M	Coldest day Date	Coldest day Mean. temp.
Alexandria Va	44.18	53.57	46.95	58	70	65	16	62.33	25	38	30	30	35.00
All Saints, S. C........	54.53	66.97	57.07	66	77	70	8	69.00	35	48	39	29	44.33
Amherst, Mass.........	34.63	43.63	37.21	51	63	56	16	53.70	15	25	21	20	25.20
Angelica, N. Y.........	33.55	41.81	33.23	52	61	54	12	54.00	10	22	18	29	19.67
Annapolis, Md ...(1)...	46.17	53.27	47.79	57	74	64	16	64.00	27	37	29	30	34.00
Ann Arbor, Mich......	....	....	....	...	...	...	...	....	...	...	...	...	....
(Woodruff)............	35.17	44.95	38.23	57	64	62	15	58.00	18	28	22	29	23.33
Ann Arbor, Mich......	....	....	....	...	...	...	...	....	...	...	...	...	....
(Winchell)............	34.83	44.27	37.24	55	65	61	15	57.67	17	29	20	29	22.47
Ashland, Va	44.60	56.23	46.53	64	76	66	11	65.67	28	40	32	22	34.33
Athens, Ill..............	38.23	52.00	42.10	63	70	62	15	63.00	18	32	24	21	25.00
Auburn, Ala.............	55.53	65.53	59.80	68	81	71	4	73.33	38	52	45	29	46.00
Augusta, Ill.............	36.17	50.00	40.33	61	71	63	14	59.33	14	31	22	21	22.67
Austin, Texas...........	50.57	70.50	55.60	74	85	72	3	76.67	35	47	44	22	45.33
Baldwinsville, N. Y....	36.90	42.67	39.90	58	58	56	12	54.33	20	22	22	29	22.67
Battle Creek, Mich....	37.13	47.43	40.00	58	65	67	15	59.67	22	31	26	29	27.00
Bedford, Pa............	38.08	47.73	41.00	62	68	54	16	59.00	17	28	25	29	26.67
Beloit, Wis...............	34.63	44.60	36.30	57	67	60	11	60.00	14	28	18	21	21.33
Beverly, N. Y...........	38.15	45.29	41.15	52	58	55	1	54.00	23	27	24	29	27.50
Blackwell's Island,....	....	....	....	...	...	...	...	....	...	...	...	...	....
N. Y.	42.55	48.53	45.03	57	65	55	16	56.33	28	30	28	29	31.00
Bladensburg, Md ..(2)..	41.04	51.05	44.10	57	63	63	3	55.33	20	37	27	29	33.67
Bloomfield, N. J........	41.10	50.75	43.35	57	67	57	2	56.83	23	29	28	29	30.16
Brandon, Vt............	32.15	41.51	31.52	52	58	56	1	52.67	15	20	10	24	17.00
Burlington, Vt.........	33.07	40.87	35.23	51	58	53	1	52.67	14	15	12	24	17.00
Burlington, N. J.......	42.23	51.73	43.40	57	70	60	2	60.33	27	31	26	29	31.33
Camden, Ark	50.30	64.87	56.30	69	81	78	4	75.67	32	48	42	22	43.00
Camden, S. C...........	50.87	66.26	56.00	65	80	70	3	70.67	30	52	32	29	40.33
Canton, N. Y...........	32.52	42.84	33.15	55	61	55	7	51.33	10	17	16	24	14.33
Carmel, Me..............	28.20	40.52	28.58	45	57	51	7	47.67	10	21	7	20	15.00
Castleton, Vt...........	32.10	43.37	34.53	56	63	56	8	52.00	12	20	11	24	18.67
Cedar Keys, Fa.........	66.03	73.03	69.37	74	79	76	9	76.33	51	58	53	29	54.33
Chapel Hill, N. C...(3)	47.79	60.73	53.08	64	81	66	,17	67.33	25	47	35	22	38.33
Charleston, S. C........	57.80	67.73	62.13	70	77	71	4	....	40	51	44	29	....
Chestertown, Md.......	42.76	49.93	44.45	55	62	58	2	58.33	28	38	28	30	32.67
Cheviot, Ohio...........	42.43	53.23	47.80	62	70	69	15	63.67	25	35	26	22	32.00
Cleveland, Ohio........	39.87	48.33	41.60	56	64	61	6	59.33	21	30	23	30	26.33
Columbus, Miss...(4)...	....	....	....	67	78	70	...	....	32	54	41	...	....
Concord, N. H.........	32.73	42.33	36.20	49	60	50	16	50.67	12	25	16	24	23.67
Cooper, Mich...........	29.25	46.57	39.60	58	66	64	15	62.67	28	30	28	22	28.00
Craftsbury, Vt.........	26.30	35.57	30.20	46	54	48	7	46.00	2	5	6	20	7.00
Crichton's Store, Va..	50.00	57.24	57.44	62	73	74	16	68.33	32	43	44	22	40·33
Danville, Ky...........	49.17	60.00	52.31	66	76	69	5	68.67	26	43	35	22	36.00
Dayton, Ohio	42.41	52.11	....	63	73	...	6	42.50	24	30	...	22	30.00
Dubuque, Iowa...(5)...	35.33	45.15	38.27	56	62	61	11	59.33	15	27	20	21	....
Deaf & Dumb Inst......	....	....	....	...	...	...	...	....	...	...	...	...	....
N. Y....................	41.44	47.90	43.79	55	59	55	1	55.16	27	29	27	29	30.33
Detroit, Mich	38.20	45.87	40.80	58	63	63	15	58.67	20	32	27	29	26·33
Delaware College,...(6)	41.12	51.14	41.16	57	66	58	2	60.00	24	32	30	29	31.67
Easton, Pa...............	37.53	49.87	40.52	54	68	56	2	56.00	19	27	27	29	27.67
Exeter, N. H...........	33.53	41.80	34.36	48	61	48	16	49.33	10	20	12	24	20.33
Flint, Mich..............	32.40	44.30	35.70	55	64	61	...	....	14	27	18	29	19.67
Flushing, N. Y...(7)...	....	....	....	...	...	...	...	....	...	...	...	...	....
Fort Madison, Iowa.(8)	35.13	48.57	40.07	61	70	61	11	62.00	13	29	20	21	21.00

(1) Obs. taken at 7½ A. M., 2 P. M. and 9 P. M. (2) Incomplete.
(3) No obs. before the third of the month—irregular till the fifth.
(4) Mean. for the month 59.02.
(5) Obs. omitted at 2 P. M. on the 21st which was probably the coldest day.
(6) Incomplete.
(7) Mean for the month 41.68, Max. 64, Min. 17.
(8) Obs. taken at 6 A. M., noon and 7 P. M.

NAME OF STATION.	Mean.			Maxima.					Minima.				
	7 A. M.	2 P. M.	9 P. M.	7 A.M.	2 P.M.	9 P.M.	Warmest day Date	Warmest day Mean temp.	7 A.M.	2 P.M.	9 P.M.	Coldest day. Date	Coldest day. Mean. temp.
Frederick City, Md	42.36	51.66	44.74	54	70	59	16	59.07	26	35	29	30	32.13
Gallipolis, Ohio	44.83	56.42	46.67	63	73	66	11	63.67	26	38	26	22	33.00
Germantown, Ohio	40.01	49.98	42.62	62	72	68	5	63.88	20	32	24	30	28.92
Gettysburg, Pa	40.20	49.80	42.10	52	72	57	3	55.33	23	32	27	30	28.00
Glenwood, Tenn	48.29	58.68	50.79	67	71	65	5	67.77	27	40	32	29	35.90
Gouverneur, N. Y	30.93	41.23	36.30	48	56	50	12	50.67	6	24	20	20	18.67
Grand Rapids, Mich	30.33	43.93	37.20	58	62	64	5	57.33	18	30	20	22	24.33
Granville, Ohio	41.67	47.80	43.13	61	68	62	15	59.67	24	31	25	29	29.67
Green Springs, Ala	55.72	69.05	60.02	69	80	72	4	73.33	37	53	41	29	44.67
Hamlin University, Min	28.67	39.50	31.58	54	54	49	9	52.37	8	21	12	21	13.76
Harrisburg, Pa	44.17	51.54	47.78	57	60	57	16	58.67	31	37	36	20	37.67
Hillsboro, Ohio	41.75	51.65	44.00	61	69	62	11	61.17	24	34	25	29	30.81
Jackson, Miss	54.23	64.47	58.50	77	79	73	16	76.33	34	50	42	22	45.00
Jackson, Ohio (Gilmor)	42.10	59.60	46.83	61	72	66	7	68.67	19	37	28	21	37.00
Jacksonville, Fa	65.36	75.93	65.43	75	82	74	8	77.00	40	54	44	29	49.33
Key West, Fa	75.63	79.23	....	80	86	...	8	83.00	62	70	...	30	66.00
Knox Hill, Fa	58.93	70.46	63.04	72	84	72	5	74.33	41	59	51	29	50.67
Lewisburg, Va	42.00	55.17	46.14	56	70	67	15	60·67	22	30	30	28	33.33
Lima, Pa	41.47	50.60	42.78	56	68	58	2	58.40	23	31	26	20	30.00
Lodi, N. Y	36.10	43.83	38.67	59	60	55	12	55.00	18	24	22	29	22.67
Londonderry, N. H	33.90	45.30	36.40	52	70	54	1	55.33	15	20	15	20	20.67
Manchester, Ill...(1)	37.80	51.27	39.80	60	72	59	11	61.00	17	30	20	21	22.33
Manchester, N. H	30.76	42.70	34.70	48	63	48	7	48.67	10	18	17	22	21.00
Meadville, Pa	37.77	47.00	39 75	59	66	57	11	59·67	18	28	25	30	25.00
Mendon, Mass	36.07	43.63	39.10	53	59	58	1	53.00	18	27	20	20	25.33
Millersburg, Ky..(2)	46.27	56.00	49.41	62	70	68	5	65.33	25	42	29	29	34.00
Milton, Ind	39.31	49.76	42.47	63	66	62	15	60.15	22	26	26	21	27.10
Milwaukie, Wis. (Winkler)	33.17	41.97	34.70	51	63	55	5	53.00	11	24	11	19	18.67
Milwaukie, Wis. (Pomeroy)	33.80	44.27	34.63	51	64	54	5	52.00	10	30	17	19	20.00
Montreal, Ca	31.32	37.81	33.48	50	54	52	8	51.55	8	12	16	24	10.00
Morrisville, Pa	40.07	50.63	42.03	54	64	55	2	56.67	24	36	27	22	32.00
Moss Grove, Pa	38.10	45.29	39.58	58	64	58	11	56.00	18	32	24	22	28.00
Muscatine, Iowa	33.36	44.80	34.90	59	67	66	5	53.67	10	28	18	21	20.00
Nantucket, Mass	45.80	42.47	45.32	58	60	58	1	57.67	28	34	28	30	34.00
Newark, N. J...(3)	....	....	....	...	...	...	16	56.12	...	...	...	23	30.37
New Bedford, Mass	39.97	47.25	41.75	52	60	57	1	56.33	22	30	21	20	26.67
Newburyport, Mass	35.10	43.30	38.45	50	65	52	16	53.70	17	22	15	20	23.40
New Harmony, Ind	43.77	54.50	47.53	65	74	67	15	68.33	22	37	30	22	31.67
New London, Conn	40.00	49.23	42.87	57	66	58	1	56.33	22	30	21	22	27.33
New Orleans, La	60.90	68.32	64.33	72	80	74	4	75.00	47	58	51	28	54.00
New Wied, Texas	55.43	73.13	59.70	78	87	79	3	80.33	40	48	44	22	47.67
Norristown, Pa	42.02	50.38	43.10	56	62	57	2	57.67	28	29	27	29	30.00
North Attleboro, Mass.	37.83	46.82	39.88	51	67	54	1	50.63	21	28	21	20	25.20
Oberlin, Ohio	40.37	49.04	41.70	58	66	61	6	60.33	21	30	23	22	31.33
Ottawa, Ill .....(4)	29.90	41.50	33.18	39	48	38	25	42.67	18	34	20	21	24.00
Oxford, Miss	51.67	62.60	54.37	66	76	70	4	67.67	34	46	40	29	42.67
Pella, Iowa	29.00	43.00	32.20	51	63	49	9	50.00	8	24	10	21	14.00
Penn Yan, N. Y..(5)	37.63	43.87	39.93	56	59	54	12	55.00	17	25	24	29	22.00

(1) Obs. taken at 7 A. M., 1 P. M. and 9 P. M.
(2) Obs. broken from the 20th to the 29th.
(3) Obs. taken at 7 A. M. and 6 P. M., Mean. for the month 44.10, Maximum 66°, Minimum 22°.
(4) For the last 11 days of the month only.
(5) Obs. taken at sunrise, 2 P. M. and sunset.

NAME OF STATION.	Mean.			Maxima.					Minima.				
							Warmest day					Coldest day.	
	7 A. M.	2 P. M.	9 P. M.	7 A.M.	2 P.M.	9 P.M.	Date	Mean temp.	7 A.M.	2 P.M.	9 P.M.	Date	Mean. temp.
Perry, Me	31.47	38.50	33.30	50	56	50	8	51.00	14	23	31	22	21.67
Perrysburg, Ohio(1)	41.13	50.73	43.50	63	70	65	15	62.67	23	31	27	29	28.00
Philadelphia, Pa	44.50	53.10	47·00	58	67	60	2	60.00	29	32	29	29	33.17
Pittsburg, Penn.	....	....	....	...	...	...	...	....	...	...	...	...	....
(Oakland Station.)	41.52	49.40	43·07	61	64	62	7	58.67	20	30	24	29	27.00
Pittsburg, Penn	....	....	....	...	...	...	...	....	...	...	...	...	....
(Marine Hospital.)(2)	43.03	49.53	45.87	62	64	60	12	60.00	24	32	27	30	30.75
Platteville, Wis	32.87	45.90	36.83	58	68	62	11	60.33	8	20	14	21	15.33
Pocopson, Pa	42.53	52.32	43.78	57	68	58	2	58.67	25	35	27	30	32.67
Pomfret, Conn	36.03	43.57	38.27	52	63	56	1	54.33	18	25	17	22	23.67
Portland, Me	33.13	41.46	35.23	49	59	52	16	49.33	14	18	13	24	21.26
Pottsville, Pa	40.85	46.54	39.02	55	66	53	12	54.10	24	27	21	29	26.33
Poultney, Iowa	28.04	42.07	31.37	52	64	56	5	50.00	8	24	8	21	14.00
Princeton, Mass	33.25	39.98	35.55	50	60	54	1	52.00	14	20	16	22	19.33
Quasqueton, Iowa..(3)	27.98	38.76	30.53	52	61	52	11	53.20	4	22	8	21	13.00
Red River Settlement.	16.04	24.46	18.88	32	40	36	2	35.33	-21	-24	-16	17	-20.33
Richmond, Mass	31.73	41.60	34.57	54	62	49	6	47.67	17	22	17	23	20.00
St. Augustine. Fa	69.90	74.30	69.50	77	82	78	5	76.67	45	59	47	29	50.67
St. James, Mich	35.60	40.65	36.43	50	55	54	13	49.00	20	21	20	28	22.00
St. Louis, Mo	42.22	52.97	46.13	62	71	67	15	66.00	24	33	27	21	28.67
St. Martins, C. E	28.87	37.63	30.12	47	57	49	7	48.80	5	8	9	20	9.67
Saugatuck, Mich	39.70	49.36	42.76	57	67	65	5	60.67	28	35	29	19	31.67
Savannah, Ohio	38.17	47.47	40.27	55	67	62	12	55.33	20	33	24	29	27.00
Savannah, Ga	57.20	69.00	61.20	71	79	70	7	71.80	38	50	44	29	45.67
Saybrook, Conn..(4)	40.89	47.96	44.56	52	60	56	1	54.67	24	30	22	22	27.33
Schellman Hall, Md	41.50	52.40	43.97	55	73	60	16	61.00	24	33	23	29	30.33
Seneca, Coll. Inst	....	....	....	...	...	...	...	....	...	...	...	...	....
Ovid, N. Y	36.67	43.33	39.43	58	60	54	12	55.33	16	22	24	29	21.33
Smithfield, Va. ..(5)	....	....	....	...	...	...	3	65.50	...	...	...	29	36.00
Smithsonian Inst	44·00	53.67	47.28	57	69	65	16	61.33	26	39	31	30	35.33
Smithville, N. Y	33.93	42.80	37.13	56	60	55	12	57.00	15	22	20	24	19.00
Sparta, Ga	51.90	66.13	56.23	68	82	70	4	70.67	34	52	34	29	40.67
Spencertown, N. Y	35.27	42.27	38.27	54	59	56	1	54.67	15	20	13	22	19.00
Springdale, Ky	43.00	55.13	51.30	64	73	71	5	68.00	19	38	33	22	30.81
Springfield, Mass	35.27	46.57	39.13	52	64	54	16	54.67	16	29	18	22	26.00
Steuben, Me	31.83	38.90	34.07	51	53	51	8	51.00	12	18	12	20	18.67
Stratford, N. H...(6)	28.93	37.87	34.07	46	54	50	8	46.33	12	15	10	24	13.33
Superior, Wis	27.04	38.40	30.24	72	57	47	28	49.33	7	21	13	21	14.33
The Rock, Ga. ...(7)	....	....	....	...	...	...	...	....	...	...	...	...	....
Unionville, Ohio	40.37	48.60	39.50	54	66	56	6	56.00	28	33	28	22	32.67
Upper Alton, Ill	38.58	51.67	44.90	65	70	66	15	67.00	18	32	24	21	26.00
Wampsville, N. Y	36.33	45.63	37.90	60	64	50	1	55.33	16	23	23	29	22.33
Warrington, Fa..(8)	62.50	70.33	67.63	73	81	75	5	76.00	45	52	56	29	54.67
Westfield, Mass	35.33	43.93	37.93	53	65	52	16	55.00	16	26	20	22	25.00
W. Haverford, Pa	42.23	51.33	....	57	68	...	16	59.00	23	30	...	29	32.00
Whitemarsh Is. Ga	57.20	67.43	61.23	70	77	70	7	71.67	37	51	44	29	45.00
Williamstown, Mass	32.90	41.39	36.83	50	58	54	1	52.50	11	21	13	29	18.67
Winchester, Va	42.87	54.67	44.87	58	72	60	16	62.67	20	36	30	30	32.33
Windham Me. ...(9)	24.09	31.32	25.77	43	41	36	15	36.33	6	15	13	30	15.33
Woodward H.	....	....	....	...	...	...	...	....	...	...	...	...	....
School, Ohio	44.67	53.47	45.03	66	72	68	15	64.33	27	34	28	22	32.67
Worcester, Mass.	35.37	45.27	38.13	50	68	58	16	53.00	16	25	17	22	22.00
Zanesville, Ohio	41.82	51.07	44.23	62	69	64	11	49.00	23	39	30	30	33.00

(1) Obs. taken at 7 A. M., 1 P. M. and 9 P. M

(2) Obs. taken at sunrise, 9 A. M., 3 and 9 P. M.—Mean. at 9 A. M. 43.80, Maximum 64°, Minimum 23°.

(3) Incomplete.

(4) For the first 25 days of the month only.

(5) Mean for the month 51.02, Max. 70, Min. 25.

(6) Obs. taken morn., noon and night, hour not specified.

(7) Mean for the month 55.04, max. 68, Min. 42—Obs taken morn., 3 P. M. and evening.

(8) Obs. taken at sunrise, noon and sunset.

(9) For the last 22 days of the month only.

NAME OF STATION.	Mean.			Maxima.					Minima.				
							Warmest day					Coldest day	
	7 A. M.	2 P. M.	9 P. M.	7 A.M.	2 P.M.	9 P.M.	Date	Mean temp.	7 A.M	2 P.M.	9 P. M	Date	Mean. temp.
Alexandria, Va	33.66	42.19	36.55	52	62	58	16	55.67	14	26	17	27	19.00
All Saints, S. C.	46.26	56.58	46.48	65	72	70	17	66.67	27	42	34	11	36.00
Ann Arbor, Mich. (Winchell)	22.70	27.37	23.39	41	53	47	1	46.13	-4	8	2	26	3.53
Ann Arbor, Mich. (Woodruff)	22.47	29.74	24.56	44	54	49	1	46.33	2	7	2	26	4.17
Angelica, N. Y.	19.74	27.52	21.94	42	44	40	2	39.67	-4	10	2	27	4.00
Annapolis, Md.	34.63	41.27	37.03	53	58	53	16	52.90	15	24	20	27	22.67
Amherst, Mass	24.41	32.44	27.52	38	46	50	2	40.50	7	11	8	29	9.00
Ashland, Va.	28.87	38.97	32.13	60	62	52	14	58.67	4	14	2	26	8.67
Athens, Ill.	21.64	33.45	25.19	49	63	53	1	52.67	-20	-2	-14	26	-7.00
Auburn, Ala.	44.51	54.90	48.30	64	67	63	23	64.33	17	32	21	26	25.33
Augusta, Ill.	18.99	30.22	22.80	49	64	50	1	51.67	-16	-3	-12	26	-6.00
Austin, Texas	51.42	58·71	47.29	68	79	66	8	70.00	7	24	16	25	17.33
Baldwinsville, N. Y.	26.26	31.10	28.35	40	50	43	2	43.67	6	10	10	29	8.67
Battle Creek, Mich.	24.67	30.81	25.68	41	61	52	1	51.00	2	6	4	30	8.67
Beloit, Wis.	18.61	26.61	21.00	47	57	48	1	50.67	-15	-6	-10	24	-9.33
Bellefontaine, Ohio.	24.02	28.23	24.97	50	46	46	2	45.00	-6	2	-2	26	-2.00
Beverly, N. Y.	28.38	33.46	30.97	47	52	49	9	46.67	12	16	11	29	13.33
Bladensburg, Md.	29.12	42.33	33.77	50	61	51	21	35.67	12	24	18	26	20.00
Bloomfield, N. J.	32.18	42.56	35.56	48	55	50	9	47.67	8	30	25	20	28.33
Boston, Mass. (1)	28.10	38.03	30.81	46	62	52	2	43.33	7	16	16	29	16.33
Brandon, Vt.	22.87	30.93	24.31	40	52	42	2	41.33	-10	3	2	29	-1.50
Burlington, Vt.	24.13	30.81	25.84	46	52	44	2	45.33	-12	5	5	29	-0.67
Burlington, N. J.	30.13	39.65	33.35	50	58	55	9	54.33	8	22	18	27	16.00
Camden, S. C.	39.58	53.74	44.52	62	72	65	16	66.33	19	36	27	11	30.67
Canton, N. Y.	22.61	28.94	24.84	46	50	50	2	45.33	-21	2	-3	29	-7.33
Carmel, Me.	16.68	29.32	20.50	38	52	46	3	39.67	-19	2	7	29	-2.00
Cedar Keys, Fa.	57.42	64.48	61.06	66	70	69	24	67.67	40	54	48	26	49.67
Chapel Hill, N. C.	37.42	51.13	41.94	59	69	63	16	63.67	22	31	25	26	27.67
Charleston, S. C.	49.19	57.48	53.00	64	68	68	17	......	32	44	40	27	......
Chestertown Md.	31.81	40.81	34.32	52	56	55	16	54.00	15	24	21	27	20.33
Cheviot, Ohio	28.13	39.71	32.94	49	61	50	1	50.00	-6	14	7	26	5.67
Cincinnati, Ohio, (Tel.) (2)	32.32	39.55	35.06	52	60	56	14	51.33	3	10	12	26	8.33
Cincinnati, O., (Lea) (3)	......	......	......	...	...	...	22	50.50	...	...	...	26	7.00
Cleveland, Ohio	27.79	33.06	28.43	50	53	47	1	45.67	1	12	2	26	9.00
Columbus, Miss. (4)	......	......	......	62	67	67	...	......	14	28	19	...	......
Concord, N. H.	22.97	32.03	26.26	37	45	44	3	38.67	-5	10	8	31	4.33
Craftsbury, Vt.	16.74	24.45	19.39	37	45	38	2	39.00	-15	3	2	29	-4.00
Crichton's Store, Va (5)	39.28	44.72	46.82	63	67	67	16	65.67	23	29	30	27	28.33
Danville, Ky.	35.84	45.94	38.16	55	65	57	14	56.33	9	19	14	26	14.00
Detroit, Mich.	26.26	31.74	27.23	45	53	47	1	46.67	3	12	6	26	8.33
Deaf & Dumb Inst. N. Y.	32.09	38.04	34.03	50	53	51	23	48.60	14	19	20	29	18.73
Dickinson College, Pa	28.16	35.67	30.02	43	53	44	23	45.50	6	18	4	31	10.67
Dubuque, Iowa	18.83	25.24	20.81	43	55	50	8	46.67	-13	-6	-9	23	9.33
Easton, Pa.	25.52	38.97	29.48	41	53	44	9	45.67	0	17	8	31	12.00
Exeter, N. H.	23.00	32.61	23.94	41	48	49	22	41.00	4	14	8	29	10.33
Flint, Mich.	21.60	30.80	23.90	44	54	46	...	......	0	12	6	26	6.00
Flushing, N. Y. (6)	......	......	......	...	...	...	...	......	...	...	...	...	......
Ft. Madison, Iowa (7)	18.39	28.94	24.74	50	60	54	8	51.00	-18	-5	-8	23	-7.00
Frederick, Md.	30.76	39.71	34.77	41	56	55	16	47.27	13	25	18	31	20.40
Germantown, Ohio.	25.68	34.45	27.72	49	52	53	14	49.17	-7	7	7	26	-2.20

(1) Obs. taken at 6 A. M., 2 P. M. and 10 P. M.
(2) Obts. taken at 6 A. M., 1 P. M. and 9 P. M. (3) Mean for the month 34.40°.
(4) Mean for the month 43.04°. (5) Obs. taken at 8½ A. M., noon, and 6½ P. M.
(6) Mean for the month 30·39, max. 50, Min. 7.5.
(7) Obs. taken at 6 A. M., noon, and 7 P. M.

NAME OF STATION.	Mean.			Maxima.					Minima.				
	7 A. M.	2 P. M.	9 P. M.	7 A.M.	2 P.M.	9 P.M.	Warmest day Date	Warmest day Mean temp.	7 A.M.	2 P.M.	9 P.M.	Coldest day. Date	Coldest day. Mean. temp.
Gettysburg, Pa.........	26.90	37.71	30.16	42	54	46	23	46.67	4	20	9	31	12.33
Glenwood, Tenn........	33.50	43.87	36.82	57	62	59	14	57.23	3	17	10	26	12.87
Gouverneur, N. Y (1).	26.93	33.60	29.60	60	56	46	9	49.33	6	15	4	29	10.87
Grand Rapids, Mich...	23.10	28.39	24.45	42	56	56	1	51.33	5	8	6	24	6.33
Granville, Ohio.........	28.74	37.84	30.77	53	64	54	1	50.67	1	13	2	26	3.67
Hamlin Univ'y, Min....	6.25	17.90	9.04	38	45	37	1	39.56	-28	-14	-22	24	-20.00
Harrisburg, Pa.........	32.55	38.71	34 77	42	55	46	23	46.67	16	23	16	31	20.00
Hillsboro, Ohio.........	28.23	36.23	30.18	52	56	52	14	47.67	-3	7	3	26	2.50
Jacksonville, Fa........	54.42	65.19	57.54	67	78	71	25	71.67	34	53	39	11	42.33
Jackson, Miss..........	42.10	51.48	46.35	64	68	74	14	64.33	12	24	20	26	21.33
Key West, Fa..........	70.26	75.97	....	78	82	...	18	80.00	54	60	...	11	57.00
Lacquiparle, Minn.....	6.35	14.13	5.97	36	42	34	7	35.33	-22	-18	-18	23	-19.33
Lewisburg, Pa.........	30.84	43.42	35.13	50	61	50	23	50.33	14	21	16	26	17.67
Lima, Pa..................	29.58	38.92	32.67	49	58	53	9	51.00	9	21	16	31	17.10
Lodi, N. Y................	25.13	32.87	27.29	40	56	40	2	43.33	5	12	9	29	11.00
Londonderry, N. H....	24.45	33.52	27.58	39	50	52	2	42.00	2	14	13	29	10.33
Manchester, Ill. (2).....	20.92	34.05	24.33	46	64	53	1	51.33	-17	0	-12	26	11.00
Manchester, N. H......	22.82	33.44	27.18	39	48	53	2	39.33	-6	13	3	31	-7.00
Meadville, Pa...........	25.35	34.74	26.90	45	50	45	1	42.67	-4	16	4	27	8.00
Mendon, Mass..........	25.84	33.74	29.29	41	52	55	9	44.00	2	17	13	27	12.33
Millersburg, Ky........	32.22	38.59	31.33	53	58	54	8	48.67	3	15	7	26	11.67
Milton, Ind...............	26.38	34.02	28.65	55	52	49	14	47.85	-8	10	0	26	0.66
Milwaukie, Wis........	....	....	....	...	...	...	...	....	...	...	...	...	....
(Winkler)	18.69	26.21	20.71	40	55	50	1	47.67	-14	-4	-8	24	-8.67
Milwaukie, Wis........	....	....	....	...	...	...	...	....	...	...	...	...	....
(Pomeroy.).	19.00	26.39	19.39	40	55	42	1	41.33	-14	-2	-8	24	-7.00
Montreal, Ca.	20.18	25.24	22.17	36	46	42	2	41.12	-14	-5	-7	29	-9.50
Morrisville, Pa.	29.28	39 13	31.94	46	54	51	23	46.00	13	19	15	29	18.67
Moss Grove, Pa........	24.00	30.71	25.63	44	52	44	1	42.67	-2	12	0	27	6.00
Muscatine, Iowa.......	17.52	27.16	19.00	46	60	52	1	45.67	-19	-5	-16	23	-9.00
Nantucket, Mass.	37.88	41.43	38.51	50	54	52	2	51.83	22	27	24	27	25.83
Nazareth, Pa............	24.74	34.45	27.24	37	52	44	9	42.67	8	17	9	31	13.00
Newark, N. J. (3)	....	....	....	...	...	...	...	49.00	...	...	...	...	18.13
New Bedford, Mass....	31.14	38.16	33.44	49	55	55	2	49.00	9	19	20	29	17.67
Newburyport, Mass....	26.27	35.02	26.92	39	50	51	23	40.90	5	14	13	30	14.00
New Harmony, Ind....	29.45	39.35	32.58	54	61	54	1	51.33	-2	10	6	26	8.00
New London, Ct........	31.13	39.68	33.74	47	57	54	23	48.67	13	20	19	29	18.33
New Orleans, La	48.77	58.33	54.02	69	72	68	15	66.83	26	35	30	25	36.67
New Wied, Texas......	43.16	60.31	50.06	70	77	70	8	71.33	13	24	20	24	23.67
Norristown, Pa.........	28.80	36.53	31.41	42	48	54	22	44.67	13	21	19	29	17.60
North Attleboro, Mass.	28.12	36.23	30.12	49	52	55	23	45.30	3	13	12	29	10.83
Oberlin, Ohio...........	28.23	34.03	29.00	49	54	47	1	46.33	5	11	7	26	9.33
Oswego, N. Y. (4)......	27.16	35.64	31.00	39	59	43	2	45.00	2	10	12	29	8.00
Ottawa, Ill................	22.25	31.67	23.66	46	63	54	2	56.33	-10	-2	-3	24	-3.00
Oxford, Miss............	36.29	47.42	39.77	57	69	58	13	59.00	8	16	11	25	14.67
Pella, Iowa.	13.48	23.87	15.90	47	54	45	8	46.33	-15	-5	-16	24	-11.33
Penn Yan, N. Y. (5) ...	25.81	33.97	29.65	39	54	46	1	45.33	8	17	13	29	13.67
Perry, Me	24.26	30.51	25.13	47	45	48	2	42.33	0	6	6	29	4.00
Perrysburg, Ohio (6)...	28.35	33.94	30.10	53	63	53	1	53.67	3	12	9	25	9.67
Philadelphia, Pa	34.00	41.30	37.20	49	60	56	16	53.00	14	21	19	27	20.00
Pittsburg, Pa............	....	....	....	...	...	...	...	....	...	...	...	...	....
(Oakland Station,)..	29.10	35.61	30.80	52	55	48	15	47.33	4	17	5	27	11.33
Pittsburg, Pa.	....	....	....	...	...	...	...	....	...	...	...	...	....
(Marine Hospital) (7)	31.03	37.19	33.13	46	56	51	15	46.50	10	17	14	27	13.00

(1) Obs. taken at Watertown, New York, after the sixth of the month

(2) Obs. taken at 7 A. M., 1 P. M. and 9 P. M.

(3) Mean for the month 33.60, max. 53½, min. 12.

(4) Obs. taken at 8 A. M., 2 P. M. and 9 P. M.

(5) Obs. taken at sunrise, 2 P. M. and sunset.

(6) Obs. taken at 7 A. M., 1 P. M. and 9 P. M.

(7) Obs. taken at sunrise, 9 A. M., 3 P. M. and 9 P. M.; Mean 9 A. M., 31.45—Max. 52°—Min. 9°.

NAME OF STATION.	Mean.			Maxima.					Minima.				
	7 A. M.	2 P. M.	9 P. M.	7 A.M.	2 P.M.	9 P.M.	Warmest day Date	Warmest day Mean temp.	7 A.M.	2 P.M.	9 P.M.	Coldest day. Date	Coldest day. Mean temp.
Platteville, Wis	15.19	23.55	16.74	42	50	48	8	45.67	-22	-10	-18	23	-14.67
Pocopson, Pa	30.20	39.94	33.21	50	58	54	9	52.00	10	22	18	27	17.67
Pomfret, Conn	26.84	33.36	27.77	45	49	53	9	42.33	7	14	14	29	12.00
Portland, Me	23 21	32.23	35.71	36	46	48	3	38.16	3	13	12	29	10.16
Pottsville, Pa. (1)	26.18	35.27	26.94	40	51	40	9	43.40	8	21	9	31	13.33
Poultney, Iowa	13.03	23.39	16.06	39	51	43	8	41.33	-18	-10	-15	23	-13.67
Princeton, Mass.	24.21	29.74	26.00	36	43	46	23	38.33	3	12	10	29	8.67
Providence, R. I	28.80	36.50	31.20	48	53	54	23	47.00	8	18	15	29	15.67
Red River Settlement	-12.21	-2.97	-9.24	22	26	22	6	22.67	-48	-30	-44	23	-39.33
Richmond, Mass	24.90	32.48	27.97	40	44	46	6	38.33	8	14	9	27	11.00
Quasqueton, Iowa	13.30	22.99	15.00	41	52	45	8	44.10	-17	-10	-11	23	-13.50
Sag Harbor, N. Y	32.90	40.84	35.74	48	56	54	16	50.33	18	22	20	27	21.00
St. James, Mich	22.88	25.59	21.98	43	45	46	1	44.67	2	3	2	24	3.67
St. Louis, Mo	27.60	37.20	31.37	54	64	56	1	54.17	-4	6	0.5	25	2.50
St. Martin, Ca., E	16.91	25.23	20.18	34	48	42	2	41.40	-23	-5	-10	29	-12.90
Saugatuck, Mich	29.22	34.35	30.71	47	59	56	1	54.00	10	15	11	24	13.67
Savannah, Ohio	26.52	35.41	28.38	50	57	42	2	46.00	0	9	3	26	6.33
Savannah, Ga	47.70	58.10	51.7	62	74	67	25	67.90	32	42	37	26	38.30
Saybrook, Ct. (2)	28.87	35.00	31.87	45	48	52	9	47.67	16	18	20	29	18.67
Schellman Hall, Md	28.45	39.97	31.81	46	63	53	23	52.00	9	19	14	27	16.67
Seneca Coll Inst., N.Y.	26.14	32.50	27.69	42	53	45	1	44.67	5	12	9	29	10.33
Smithsonian Inst	33.13	41.44	36.74	49	58	57	16	53.67	17	25	19	27	22.33
Smithville, N. Y	23.81	30.26	27.03	48	48	48	2	44.67	-13	4	5	29	-1.33
Smithfield, Va. (3)	....	....	....	...	...	...	15	62.00	...	...	...	27	27.50
Sparta, Ga	40.81	53.65	44.84	63	69	63	23	64.00	20	34	30	26	28.33
Spencertown, N. Y	24.10	31.94	27.45	45	49	47	2	43.33	-3	9	8	31	6.33
Springdale, Ky	29.18	40.42	35.76	53	61	56	14	55.00	-6	16	11	26	7.00
Springfield, Mass	25.16	35.61	27.81	42	52	48	23	42.00	2	15	9	29	10.33
Steuben, Me	23.29	32.77	25.42	46	50	46	2	44.67	-3	10	2	29	5.33
Stratford, N. H. (4)	19.68	27.84	25.26	39	44	43	2	38.00	-12	8	6	29	0.67
Superior, Wis	8.35	18.40	11.30	36	42	32	1	37.67	-25	-10	-19	24	-17.00
The Rock, Ga. (5)	....	....	....	...	...	...	...	....	...	...	...	...	....
Unionville, Ohio	28.65	33.68	28.97	46	52	40	1	43.33	6	14	6	26	11.33
Upper Alton, Ill	24.42	36.00	29.57	50	62	56	1	52.67	-8	3	-5	24	0.33
Urbana, Ohio	25.13	34.16	27.13	49	54	50	1	46.00	-5	5	-2	26	0.67
Wampsville, N. Y	25.58	34.29	29.90	40	56	44	2	46.00	0	13	10	29	7.67
Warrington, Fa	53.06	59.45	58.03	72	78	74	16	72.67	23	39	38	26	33.33
Westfield, Mass	24.42	33.29	27.74	41	50	45	23	40.33	-5	12	4	31	7.00
West Haverford, Pa	31.97	40.71	....	48	59	...	9	52.00	13	20	...	31	20.00
Whitemarsh Island, Ga	46.97	57.65	52.26	63	68	65	25	65.33	30	43	39	26	37.33
Williamstown, Mass	23.69	30.32	26.18	45	47	45	2	40.30	3	9	6	29	8.90
Winchester, Va	30.42	42.03	33.52	42	60	50	23	50.33	10	24	14	31	18.67
Windham, Me	24.15	28.90	23.40	33	36	33	3	33.33	6	10	8	29	8.67
Woodward High School	30.16	40.55	33.39	50	55	55	14	51.00	-2	19	8	26	10.67
Worcester, Mass	25.84	35.42	28.52	45	57	53	23	42.67	5	12	10	29	9.67
Zanesville, Ohio	28.13	38.68	30.97	52	57	50	22	49.33	2	14	5	26	8.00

(1) But few obs. on the Thermometer after the 9th of the month.
(2) For the last 23 days of the month.
(3) Monthly mean 41.27, max. 64, min. 19.
(4) Obs. taken morning, noon and night—hours not specified.
(5) Mean for the month 45.14, maximum 64, minimum 26—obvs. taken morning, 3 P. M. and evening.

NAME OF STATION.	Mean force of Vapor.			Relative Humidity.								
				Mean.			Maxima.			Minima.		
	7 A. M.	2 P. M.	9 P. M.	7 A. M.	2 P. M.	9 P. M.	7 A. M.	2 P. M.	9 P. M.	7 A. M.	2 P. M.	9 P. M.
Aiken, S. C.	.228	.281	.293	80	58	77	100	92	100	58	00	43
Alexandria, Va.	.163	.202	.177	89	76	87	100	100	100	71	47	66
All Saints, S. C.	.251	.290	.298	82	59	83	100	93	100	45	16	40
Amherst, Mass	.133	.150	.135	93	85	91	100	100	100	74	54	61
Ann Arbor Mich.	...	...	...	...	...	...	...	...	...	...	...	...
(Winchell).	.116	.130	.128	90	79	87	100	100	100	70	52	69
Auburn, Ala,	.233	.272	.239	75	53	65	100	87	94	28	08	32
Augusta, Ga.	.251	.314	.276	83	72	75	100	100	100	31	36	10
Austin Texas.	.240	.247	.277	83	47	80	100	100	100	45	00	34
Bloomfield, N. J.	.148	.157	.149	85	72	84	100	93	95	45	42	57
Burlington, Vt.	.095	.108	.095	68	60	69	84	88	90	00	00	46
Camden, S. C.	.214	.270	.246	87	59	78	100	100	93	53	31	49
Charleston, S. C.(1)	...	...	...	...	...	...	...	...	...	...	...	...
Deaf and Dumb Inst.	...	...	...	...	...	...	...	...	...	...	...	...
N. Y.	.144	.150	.144	76	77	80	100	100	100	39	33	55
Dubuque, Iowa.	.114	.119	.117	89	70	85	100	100	100	58	11	39
Frederick, Md.	.151	.170	.162	87	71	82	100	100	100	61	39	47
Gallipolis, Ohio.	.163	.222	.187	84	74	81	92	100	100	76	44	57
Germantown, Ohio.	.135	.157	.143	90	72	82	98	96	97	50	43	40
Glenwood, Tenn.	.187	.192	.202	80	58	77	100	99	100	41	14	44
Hillsboro, Ohio.	.156	.185	.165	90	82	89	100	100	100	56	41	58
Jacksonville, Fa.	.331	.446	.375	86	74	89	100	94	94	58	39	80
Knoxville, Tenn.	.199	.218	.231	88	68	85	100	100	100	62	14	46
Lima, Pa.	.156	.163	.157	85	70	85	100	100	100	52	35	34
Madison, Ind.	.168	.220	.184	89	84	86	100	97	100	64	60	63
Nanchester, Ill.	.146	.192	.160	94	88	94	100	100	100	63	56	60
Nantucket, Mass.	.168	.168	.175	76	65	77	100	92	100	46	33	45
New Bedford, Mass.	.159	.186	.164	95	82	91	100	96	100	84	62	15
Newburyport, Mass.	.122	.134	.128	65	61	66	90	100	88	04	04	15
New Harmony, Ind.	.176	.204	.203	93	79	94	100	100	100	55	46	63
New Orleans, La.	.331	.341	.355	83	60	78	100	100	100	10	26	27
New Wied, Texas.	.268	.366	.279	78	61	76	100	94	100	37	00	00
Norristown, Pa.	.134	.150	.127	77	69	70	100	93	100	36	44	44
North Attleboro, Mass.	.147	.162	.147	83	74	85	100	100	100	40	38	44
Oxford, Miss.	.227	.258	.239	87	70	78	100	100	100	55	28	36
Penn Yan, N. Y	...	.132	...	...	72	...	...	100	...	...	50	...
Philadelphia, Pa.	.159	.167	.162	84	72	79	96	96	100	66	36	54
Pomfret, Conn.	.138	.173	.141	93	89	93	100	100	100	78	62	76
Princeton, Mass.	.110	.119	.116	81	83	78	100	100	100	00	38	00
Sacramento, Cal.	...	...	...	...	...	...	...	...	...	...	...	...
(Logan.) (2)	...	...	...	...	...	...	...	...	...	...	...	...
Sacramento, Cal.	...	...	...	...	...	...	...	...	...	...	...	...
(Hatch.) (3)	...	...	...	...	...	...	...	...	...	...	...	...
St. Louis, Mo.	.144	.169	.157	77	63	73	95	95	93	49	23	39
St. Martin, Ca., E.	.084	.125	.100	91	89	92	98	99	99	34	59	69
Savannah, Ga.	.266	.286	.299	79	53	74	96	92	96	46	18	35
Smithsonian Inst.	.168	.188	.189	88	72	87	100	100	100	40	36	58
Springdale, Ky.	.157	.201	.195	85	75	81	100	100	100	54	34	49
Thornbury, N. C.	.178	.199	.201	80	50	72	95	84	95	38	19	29
Tuscaloosa, Ala.	.242	.326	.262	83	54	73	96	88	94	49	26	36
Upper Alton, Ill.(4)	.253	.373	.286	89	92	86	100	100	100	73	85	72
Westfield, Mass.	.119	.138	.117	79	77	76	100	100	100	40	12	34
Williamstown, Mass.	.130	.154	.129	91	87	87	100	100	100	57	44	36
Worcester, Mass.	.092	.131	.097	54	73	61	100	100	100	00	00	00
Zanesville, Ohio.	.138	.176	.152	78	71	80	100	100	100	35	30	38

(1) Mean force of vapor for the month 292, computed approximately from the mean dew point.
(2) Mean force of vapor 230, computed approximately from the mean dew points for the month, maximum 294, minimum 167. (3) Obs. taken at sunrise, 3 P. M. and 10 P. M., mean force of vapor 221, maximum 336, minimum 134, mean humidity 73, max. 89, min. 55. The means are computed approximately from the mean dew point for each day. (4) Incomplete.

NAME OF STATION.	Mean force of Vapor.			Relative Humidity.								
				Mean.			Maxima.			Minima.		
	7 A. M.	2 P. M.	9 P. M.	7 A. M.	2 P. M.	9 P. M.	7 A. M.	2 P. M.	9 P. M.	7 A. M.	2 P. M.	9 P. M.
Aiken, S. C	.184	.220	.213	77	52	69	100	100	100	48	19	40
Alexandria, Va	.069	.152	.131	81	82	79	100	100	100	26	58	63
All Saints, S. C	.199	·226	.228	71	57	79	94	100	94	50	24	28
Amherst, Mass	.095	.114	.104	95	81	93	100	100	100	81	50	68
Ann Arbor, Mich	......	......	......	......	......	......	......	......	......	...	...	...
(Winchell)	.081	.097	.083	96	83	92	100	100	100	77	59	64
Auburn, Ala	.158	.159	.167	68	41	57	93	77	89	40	08	22
Angusta, Ga......(1)	.230	.322	.248	80	80	73	100	94	85	10	50	32
Austin, Texas	.220	.263	.286	90	54	85	100	100	100	58	18	48
Bloomfield, N. J	.093	.112	.105	79	66	77	95	95	92	57	28	17
Burlington, Vt	.064	.077	.071	62	50	60	100	87	89	00	00	00
Camden, S. C	.158	.182	.180	84	47	71	100	94	94	13	10	38
Charleston, S. C..(2)	......	......	......	......	......	......	......	......	......	...	...	...
Dubuque, Iowa	.079	.102	.090	90	74	88	100	100	100	66	44	62
Deaf &Dumb Inst.,N Y	.098	.113	.100	77	72	70	100	100	97	36	35	10
Frederick, Md	.107	.120	.113	81	66	74	100	90	100	64	39	41
Gallipolis, Ohio	.107	.163	.125	82	81	80	94	100	100	64	00	53
Germantown, Ohio	.094	.110	.102	91	71	83	97	98	98	77	49	51
Glenwood, Tenn	.136	.133	.144	83	54	67	98	100	90	42	26	25
Green Springs, Ala	.200	.237	.198	81	54	69	94	89	94	58	07	34
Hillsboro, Ohio	.108	.130	.136	100	84	97	100	100	100	87	47	67
Jacksonville, Fa	.282	.380	.320	86	70	88	94	94	100	66	51	71
Knoxville, Tenn	.139	.149	.135	78	51	67	98	92	88	37	25	28
Lima, Pa	.079	.105	.106	78	62	75	100	100	100	46	23	44
Madison, Ind	.130	.154	.148	88	79	87	100	93	100	74	52	65
Manchester, Ill	.108	.138	.135	93	85	100	100	100	100	31	49	100
Nantucket, Mass	.113	.106	.120	69	54	74	96	100	100	00	00	59
New Bedford, Mass	.100	.126	.105	85	75	83	100	100	100	59	51	60
Newburyport, Mass	.075	.084	.085	59	54	61	100	100	100	11	15	04
New Harmony, Ind	.131	.154	.148	95	68	94	100	100	100	61	30	70
New Orleans, La	·266	.264	.301	86	57	81	100	92	100	58	31	46
New Wied, Texas	.240	.341	.319	79	62	87	100	100	100	25	13	28
Norristown, Pa	.087	.099	.090	62	53	59	100	100	100	00	02	00
North Attleboro, Mass.	.080	.093	.091	80	62	77	100	100	100	00	30	47
Oxford, Miss	.160	.178	.183	85	57	73	100	100	100	61	30	40
Penn Yan, N. Y	......	.119	......	...	68	...	.....	97	.....	...	60	...
Philadelphia, Pa	.102	.116	.113	77	64	72	100	100	95	45	23	44
Pomfret, Conn	.100	.122	.099	95	87	90	100	100	100	76	53	76
Princeton, Mass	.079	.083	.068	84	71	65	100	100	100	00	26	14
Sacramento, Cal	......	......	......	......	......	......	......	......	......	...	...	...
(Logan). ...(3)	......	......	......	......	......	......	......	......	......	...	...	...
do. do. (Hatch, (4)	......	......	......	......	......	......	......	......	......	...	...	...
St. Louis, Mo	.114	.121	.127	80	57	74	93	88	92	55	25	47
St. Martin, Ca., East	.063	.102	.078	85	85	88	99	98	100	22	63	46
Savannah, Ga	.209	.203	.221	75	44	63	93	89	89	43	16	29
Smithsonian, Inst	.113	.112	.103	78	60	77	100	100	100	38	37	46
Springdale, Ky	.114	.154	.152	85	79	84	100	100	100	60	42	62
Thornbury, N. C	.130	.118	.127	78	40	66	100	77	92	37	17	35
Tuscaloosa, Ala	.182	.180	.179	79	46	66	91	92	92	56	12	43
Upper Alton, Ill..(5)	.217	.222	.252	84	94	78	85	100	82	84	81	75
Westfield, Mass	.075	.089	.085	65	64	67	100	91	100	00	35	04
Williamstown, Mass	.093	.103	.093	94	86	94	100	100	100	60	63	74
Woodward H. School.	.080	.108	.110	68	60	68	93	87	100	41	10	29
Worcester, Mass	.139	.220	.181	45	56	49	100	100	100	00	00	00

(1) For the first 14 days of month only. (2) Mean force of vapor 224, computed approximately from the mean dew point. (3) Mean force of vapor 261. (computed approximately from the mean dew point for the month)—Maximum, 381—Minimum, 098. (4) Obs. taken at sunrise, 3 P. M. and 10 P. M. Mean force of vapor, 266—Maximum, 437—Minimum, 102. Mean humidity, 66; Max. 92, Min. 30. The means are computed approximately from the mean dew point for each day.
(5) Month incomplete.

NAME OF STATION.	Mean force of Vapor.			Relative Humidity.								
				Mean.			Maxima.			Minima.		
	7 A. M.	2 P. M	9 P. M.	7 A. M.	2 P. M.	9 P. M.	7 A. M.	2 P. M.	9 P. M.	7 A. M.	2 P. M.	9 P.M.
Aiken, S, C.	.260	.246	.269	74	41	59	100	89	94	47	09	19
Alexandria, Va.	.181	.215	.209	80	68	82	100	94	100	30	39	56
All Saints, S. C.	.290	.309	.307	82	53	75	100	100	94	49	18	17
Amherst, Mass.	.133	.139	.136	86	69	81	100	100	100	63	33	62
Ann Arbor, Mich.	......	......	......	......	......	......	......	......	......	...	...	...
(Winchell)	.115	.129	.122	81	67	77	100	100	100	38	37	38
Auburn, Ala.	.236	.220	.244	64	32	50	100	85	92	28	07	15
Augusta, Ga...(1)	.333	.325	.264	86	68	66	100	100	100	56	26	15
Austin, Texas,	.286	.263	.375	84	37	84	100	100	100	43	00	42
Bloomfield, N. J.	.118	.151	.147	67	58	72	90	91	95	36	22	41
Burlington, Vt.	.088	.101	.103	68	51	67	87	100	100	43	04	46
Camden, S. C.	.249	.267	.266	79	46	65	100	84	92	50	11	35
Charleston, S. C...(2).	......	......	......	......	......	......	......	......	......	...	...	...
Deaf & Dumb Inst., N.Y	.135	.152	.140	72	66	69	100	100	100	15	35	29
Dubuque, Iowa.	.127	.157	.150	82	69	84	100	100	100	60	31	60
Frederick City, Md.	.139	.159	.152	71	56	65	98	95	99	23	00	36
Gallipolis, Ohio.	.181	.205	.181	81	62	73	100	92	92	62	15	45
Germantown, Ohio.	1.39	.167	.144	82	54	76	98	98	100	48	26	46
Glenwood, Tenn.	.172	.173	.201	70	42	62	94	86	98	24	14	29
Green Springs Ala.	.260	.311	.282	72	47	60	100	85	100	31	16	21
Hillsborough, Ohio.	.162	.177	.164	89	69	77	100	100	100	49	30	40
Jacksonville, Fa.	.381	.523	.402	81	69	82	100	94	94	00	44	49
Knoxville, Tenn.	.185	.197	.199	77	41	60	100	88	88	33	16	00
Lebanon, Tenn.	.161	.176	.194	55	44	52	100	75	90	00	20	00
Lima, Pa.	.137	.136	.146	71	52	69	100	100	100	31	24	35
Madison, Ind.	.180	.219	.202	81	75	79	100	91	96	67	51	52
Manchester, Ill.	.158	.203	.194	90	77	93	100	100	100	06	52	68
Muscatine, Iowa...(1).	.188	.274	.161	100	100	100	100	100	100	100	100	100
Nantucket, Mass.	.141	.139	.161	68	56	78	100	89	95	45	32	46
New Bedford, Mass.	.138	.173	.150	79	68	84	100	95	100	53	51	66
Newburryport, Mass.	.122	.131	.134	70	53	72	100	96	100	52	25	32
New Harmony, Ind.	.150	.197	.205	89	60	88	100	100	100	52	32	39
New Orleans, La.	.368	.343	.367	84	52	72	100	80	100	48	16	35
New Wied, Texas...(1)	.297	.569	.435	72	60	84	100	89	100	28	24	33
Norristown, Pa.	.149	.182	.166	76	67	75	100	100	100	08	14	44
North Attleboro, Mass.	.108	.113	.115	68	48	67	100	100	100	47	07	39
Oxford, Miss...(1)	.242	.318	.272	78	55	70	100	100	100	12	31	34
Penn Yan, N.Y.	......	.125	......	...	61	...	.....	100	.....	...	29	...
Philadelphia, Pa	.143	.152	.162	71	53	65	94	92	91	47	27	43
Pomfret, Conn.	.124	.155	.134	83	70	81	100	100	100	52	42	57
Pottsville, Pa ...(1)	.097	.132	.096	63	57	58	66	100	79	60	15	37
Princeton, Mass.	.096	.103	.093	68	56	62	100	98	100	43	24	24
Sacramento, Cal.	......	......	......	.....	......	......	......	......	......	...	...	...
(Logan). (3)	......	......	......	......	......	......	......	......	......	...	...	...
do. do (Hatch)(4)	......	......	......	......	......	......	......	......	......	...	...	...
St. Louis, Mo.	.146	.168	.171	71	52	65	100	90	95	44	20	37
St. Martin, Ca., E.	.091	.156	.111	89	82	84	98	97	98	66	38	26
St. Martin, Ca...(5)	.102	.167	.107	85	79	82	95	96	96	70	56	63
Savannah, Ga.	.298	.271	.312	74	40	58	93	93	93	37	11	24
Smithsonian Inst.	.166	.159	.191	72	51	75	100	100	100	24	11	34
Springdale, Ky.	.162	.210	.206	84	72	78	100	100	100	59	37	50
Thornbury, N. C.	.208	.187	.205	76	42	62	95	85	96	43	12	29
Tuscaloosa, Ala.	.248	.234	.260	72	36	58	89	84	88	42	12	25
Upper Alton, Ill..(1).	.226	.354	.267	91	99	82	100	100	94	80	61	41
Warrington, Fa.	.411	.507	......	100	84	...	100	100	...	100	56	...
Westfield, Mass.	.069	.112	.115	66	50	65	100	100	100	38	24	30
Williamstown, Mass.	.131	.154	.140	92	82	87	100	100	100	66	48	64
Worcester, Mass.	.099	.139	.116	62	64	63	100	100	100	05	20	11

(1) Incomplete. (2) Mean force of vapour 295, computed approximately from the mean dew point.
(3) Mean force of vapour 301, (computed approximately from the mean dew point for the month,) Maximum, 500—Minimum, 181. (4) Obs. taken at sunrise, 3 P. M. and 10 P. M.; Mean force of vapour, 342—Max., 478—Min., 120—Mean humidity, 73—Max., 93 The means are computed approximately from the mean dew point for each. (5) The values in this line are computed from the same observations as the preceeding, by using the reductions of the observer, who employed for the purpose a different set of tables.

NAME OF STATION.	Mean force of Vapor.			Relative Humidity. Mean.			Maxima.			Minima.		
	7 A. M.	2 P. M.	9 P. M.	7 A. M.	2 P. M.	9 P. M.	7 A. M.	2 P. M.	9 P. M.	7 A. M.	2 P. M.	9 P. M.
Aiken, S. C	.364	.384	.381	66	43	58	88	89	100	34	19	28
Alexandria, Va	.313	.397	.354	84	66	77	100	100	100	33	18	44
All Saints, S. C	.431	.452	.447	84	61	78	100	90	95	59	21	59
Amherst, Mass	.211	.235	.217	87	65	81	100	100	100	53	26	49
Ann Arbor, Mich	.....	.....	.....	...	...	...	.....	...	.....	...	..	...
(Winchell)	.234	.241	.248	73	49	68	100	80	98	46	22	37
Auburn, Ala	.358	.310	.328	64	33	48	100	88	100	21	13	15
Austin, Texas	.505	.443	.505	89	41	75	100	100	100	66	22	41
Bloomfield, N. J	.203	.268	.263	73	61	77	95	94	97	12	19	47
Burlington, Vt	.171	.188	.173	67	48	63	100	73	91	42	09	40
Camden, S. C	.367	.366	.405	73	41	65	94	75	88	34	19	30
Charleston, S. C.(1)	.....	.....	.....	...	...	...	.....	.....	.....	...	...	...
Deaf & Dumb Inst., N Y	.218	.237	.218	75	61	70	100	100	100	36	22	18
Dubuque, Iowa	.267	.249	.297	71	39	62	100	86	100	37	12	28
Frederick, Md	.250	.280	.280	71	46	65	100	100	97	32	22	37
Gallipolis, Ohio	.300	.382	.331	69	51	67	93	91	87	43	13	48
Germantown, Ohio	.277	.269	.297	72	40	73	98	93	100	26	18	35
Glenwood, Tenn	.321	.313	.329	72	41	60	99	93	98	45	19	28
Green Springs, Ala	.486	.554	.484	75	52	68	94	100	94	39	36	36
Hillsboro, Ohio	.279	.274	.266	73	43	59	100	100	100	44	04	31
Jacksonville, Fa	.542	.664	.583	82	70	88	95	90	95	15	52	62
Knoxville, Tenn	.312	.264	.307	68	39	59	97	100	95	26	00	24
Lima, Pa	.254	.261	.255	76	52	73	97	100	97	39	12	39
Madison, Ind	.316	.352	.360	75	52	69	96	90	92	45	24	49
Manchester, Ill	.312	.474	.355	79	61	74	100	100	100	58	34	43
Muscatine, Iowa	.256	.370	.305	72	51	73	100	88	100	34	28	32
Nantucket, Mass	.196	.193	.216	68	54	80	100	95	96	22	35	20
New Bedford, Mass	.221	.276	.248	84	70	82	96	95	100	71	43	77
Newburyport, Mass	.172	.191	.191	69	57	73	100	100	100	32	15	36
New Harmony, Ind	.319	.342	.379	75	47	75	100	92	100	42	26	41
New Orleans, La	.526	.491	.567	88	51	78	100	77	100	45	24	49
New Wied, Texas	.578	.776	.587	80	67	83	100	100	100	75	20	60
Norristown, Pa	.271	.328	.304	80	72	80	100	100	100	37	22	42
North Attleboro, Mass.	.172	.165	.181	64	46	66	94	81	91	20	02	24
Ottawa, Ill	.333	.618	.385	90	86	94	100	100	100	47	65	79
Oxford, Miss	.415	.463	.401	69	53	61	100	100	100	30	19	18
Penn Yan, N. Y	.....	.238	.....	...	56	...	.....	88	...	...	21	...
Philadelphia, Pa	.246	.281	.281	72	54	67	91	93	96	44	22	34
Pomfret, Conn	.209	.262	.217	86	69	84	100	100	100	57	41	48
Pottsville, Pa	.193	.235	.229	68	52	72	100	100	100	35	09	29
Princeton, Mass	.155	.163	.165	66	53	71	100	100	100	00	13	36
St. Louis, Mo	.259	.292	.292	53	40	53	89	84	93	31	17	32
St. Martin, Ca., East	.160	.241	.190	81	72	75	100	100	95	46	47	55
Savannah, Ga	.441	.440	.471	77	48	75	96	92	91	55	25	49
Smithsonian Inst	.271	.302	.297	75	49	66	100	100	97	26	14	37
Spencertown, N. Y	.202	.247	.181	77	69	67	100	100	92	45	35	06
Thornbury, N. C	.277	.291	.304	73	36	62	91	90	82	27	11	33
Upper Alton, Ill	.368	.607	.386	78	74	70	100	100	100	02	44	25
Warrington, Fa	.693	.699	.....	98	90	...	100	100	...	68	70	...
Westfield, Mass	.168	.214	.184	71	55	65	100	100	92	40	15	35
Williamstown, Mass	.513	.625	.625	87	73	86	100	100	100	39	17	66
Worcester, Mass	.168	.214	.197	69	59	72	100	100	100	25	24	00

(1) Mean force of vapor, 491, computed approximately from the mean dew point.

NAME OF STATION.	Mean force of Vapor.			Relative Humidity. Mean.			Maxima.			Minima.		
	7 A. M.	2 P. M.	9 P. M.	7 A. M.	2 P. M.	9 P. M.	7 A. M.	2 P. M.	9 P. M.	7 A. M.	2 P. M.	9 P. M.
Aiken, S. C.	.508	.548	.500	77	55	62	100	84	100	54	34	44
Alexandria, Va	.385	.451	.415	75	57	73	100	100	100	47	32	54
All Saints, S. C.	.555	.572	.565	81	63	78	91	95	95	48	32	46
Amherst, Mass	.300	.287	.292	81	44	71	100	74	100	56	22	39
Ann Arbor. Mich	......	......	......	...	...	...	.....	.....	.....	...	...	...
(Winchell).	.264	.275	.265	59	40	60	97	96	90	28	18	32
Auburn, Ala,	.501	.397	.440	69	34	55	94	65	95	49	16	28
Augusta, Ga. (1)	.541	.600	.618	71	60	83	90	81	100	82	47	59
Austin, Texas	.608	.549	.605	88	43	68	100	72	100	56	27	52
Bloomfield, N. J.	.296	.361	.340	70	52	75	93	92	97	40	22	33
Burlington, Vt	.225	.286	.231	57	43	52	100	79	71	24	23	31
Camden, S. C	.491	.502	.502	74	50	68	90	100	95	49	31	31
Charleston, S. C. (2)	......	......	......	...	...	...	.....	.....	.....	...	...	...
Deaf & Dumb Inst. N. Y.	.305	.305	.308	68	50	67	100	100	100	43	25	31
Dubuque, Iowa	.314	.346	.336	62	45	58	100	93	100	28	21	16
Frederick, Md	.313	.339	.345	62	41	61	99	85	94	33	18	30
Gallipolis, Ohio	.419	.500	.534	73	56	74	89	79	100	53	24	43
Gardiner, Me	.313	.339	......	80	61	...	100	100	.....	36	45	...
Germantown, Ohio	.365	.366	.377	74	45	77	95	96	95	36	20	40
Glenwood, Tenn	.439	.473	.483	84	54	78	95	96	98	46	21	58
Green Springs, Ala	.598	.601	.621	76	45	68	100	78	86	54	22	53
Hillsboro, Ohio	.381	.427	.361	81	56	69	100	100	100	58	28	40
Jacksonville, Fa.	.681	.791	.713	84	67	88	91	83	91	59	45	61
Knoxville, Tenn	.434	.430	.440	76	47	70	90	87	91	53	17	48
Lima, Pa.	.310	.279	.321	70	42	67	97	95	93	39	10	36
Madison, Ind	.434	.481	.474	77	56	71	91	88	90	43	35	46
Madison, Wis	.266	.268	.250	45	39	40	89	70	94	23	09	10
Manchester, Ill	.430	.573	.441	82	67	82	100	90	100	64	43	48
Muscatine, Iowa	.332	.476	.359	78	61	73	100	100	94	53	32	45
Nantucket, Mass	.271	.252	.288	71	55	82	94	88	100	39	14	56
New Bedford, Mass	.332	.373	.330	87	67	91	100	100	100	68	26	73
Newburyport, Mass	.258	.278	.285	69	52	73	100	97	98	35	19	23
New Harmony, Ind	.444	.492	.495	79	55	82	100	90	100	49	13	61
New Orleans, La	.669	.641	.705	89	53	79	95	95	98	53	39	64
New Wied, Texas(1)	.732	.893	.684	85	58	82	100	100	100	76	33	66
Norristown, Pa.	.285	.295	.285	74	47	62	93	78	89	45	12	08
North Attleboro, Mass.	.243	.223	.240	62	38	70	92	96	97	25	13	28
Ottawa, Ill	.466	.727	.448	87	85	91	100	100	100	54	44	52
Oxford, Miss.	.600	.624	.546	67	61	73	100	100	100	37	33	20
Penn Yan, N. Y	......	.275	......	...	48	...	.....	89	.....	...	14	...
Philadelphia, Pa	.326	.321	.340	68	43	60	97	100	93	41	16	41
Pomfret, Conn	.307	.373	.300	84	63	83	100	91	100	58	43	50
Pottsville, Pa	.240	.276	.274	63	42	66	93	100	100	32	16	32
Princeton, Mass	.213	.228	.234	62	45	63	100	94	98	32	09	22
St. Louis, Mo	.416	.463	.448	71	54	71	93	90	93	44	09	50
St. Martin, Ca., E.	.257	.403	.286	78	60	71	98	81	94	54	28	41
Savannah, Ga	.582	.547	.586	66	47	74	91	75	92	42	30	53
Smithsonian Inst	.376	.372	.403	78	46	68	100	100	100	44	21	39
Spencertown, N. Y.	.272	.303	.247	72	52	73	100	100	100	52	22	38
Springdale, Ky	.390	.587	.515	92	65	78	100	100	95	76	34	41
Upper Alton, Ill	.465	.660	.512	89	82	83	100	100	100	75	58	54
Westfield, Mass	.243	.255	.272	68	38	64	93	65	100	47	17	34
Williamstown, Mass(1)	.286	.310	.302	85	64	74	100	85	100	64	40	49
Worcester, Mass	.239	.270	.261	62	46	67	100	100	100	25	05	39

(1) Incomplete.
(2) Mean force of vapor, 632, computed approximately from the mean dew point.

NAME OF STATION.	Mean force of Vapor.			Relative Humidity.								
				Mean.			Maxima.			Minima.		
	7 A. M.	2 P. M	9 P. M.	7 A. M.	2 P. M.	9 P. M.	7 A. M.	2 P. M.	9 P. M.	7 A. M	2 P. M.	9 P. M.
Aiken, S, C.	.627	.634	.626	80	58	73	100	81	95	61	37	44
Alexandria, Va.	.582	.656	.609	82	69	85	100	100	100	58	42	68
All Saints, S. C.	.684	.718	.702	82	74	79	95	95	95	66	50	45
Amherst, Mass.	.493	.513	.496	89	65	86	100	100	100	50	36	61
Ann Arbor, Mich.	.....	......	......	...	...	...	.....	.....	.....	...	...	...
(Winchell)	.431	.442	.428	78	61	85	99	99	100	42	25	51
Auburn, Ala.	.562	.488	.554	72	42	64	95	86	90	49	16	29
Augusta, Ga... (1)	.752	.895	.851	86	83	82	100	91	100	67	43	59
Austin, Texas,	.600	.564	.624	75	51	68	100	73	100	51	31	49
Bloomfield, N. J.	.482	.560	.517	83	64	85	97	95	97	58	35	52
Burlington, Vt.	.425	.507	.421	76	64	75	95	100	100	57	35	54
Camden, S. C.	.603	.631	.621	75	56	78	95	100	90	32	37	62
Charleston, S. C... (2).	......	......	......	...	..	...	...	...	...	...	...	...
Deaf & Dumb Inst., N.Y	.485	.497	.494	75	61	75	98	91	100	51	24	44
Dubuque, Iowa.	.459	.478	.481	73	53	74	92	100	100	53	20	52
Frederick, Md.	.516	.555	.543	72	59	79	100	100	100	51	27	38
Gallipolis, Ohio.	.554	.634	.640	82	71	78	90	94	100	68	56	42
Gardiner, Me.	.553	.630	......	92	79	...	100	100	...	71	47	...
Germantown, Ohio.	.531	.550	.522	85	63	87	97	96	98	67	44	61
Glenwood, Tenn.	.557	.568	.592	90	61	85	96	94	98	74	33	64
Green Springs, Ala.	.668	.690	.650	79	55	76	95	79	91	58	34	54
Hillsboro, Ohio.	.533	.587	.529	95	74	90	100	100	100	75	46	65
Jacksonville, Fa.	.821	.889	.802	89	79	90	100	100	100	71	50	81
Knoxville, Tenn.	.517	.507	.519	78	52	76	93	93	100	65	27	53
Lima, Pa.	.525	.549	.545	82	63	89	98	95	97	60	39	69
Madison, Ind.	.549	.638	.598	87	73	82	95	88	92	66	53	55
Manchester, Ill.	.540	.647	.569	81	65	85	100	95	100	38	46	64
Muscatine, Iowa.	.531	.653	.536	85	69	86	100	100	100	50	39	57
Nantucket, Mass.	.425	.451	.444	72	65	86	92	84	97	29	42	75
New Bedford, Mass (1)	.527	.549	.542	87	73	91	100	97	100	54	46	70
Newburyport, Mass.	.437	.444	.430	76	57	75	95	93	100	43	24	50
New Harmony, Ind.	.605	.604	.606	82	62	84	100	90	95	58	45	66
New Orleans, La.	.737	.739	.758	88	62	82	100	100	100	51	30	35
New Wied, Texas.	.777	.922	.775	83	63	89	100	92	100	60	29	60
Norristown, Pa.	.459	.489	.482	77	59	72	94	84	100	42	32	36
North Attleboro, Mass.	.434	.431	.441	73	55	80	93	91	96	43	18	59
Ottawa, Ill.	.633	.815	.615	94	90	95	100	100	100	79	77	87
Oxford, Miss.	.639	.642	.651	84	63	84	100	100	100	41	40	59
Penn Yan, N.Y.	......	.400	......	...	53	...	...	93	...	...	20	...
Philadelphia, Pa	.523	.537	.576	75	57	73	94	90	95	55	34	53
Pomfret, Conn.	.488	.555	.494	89	76	89	100	100	100	70	53	50
Pottsville, Pa	.422	.453	.471	73	53	79	100	94	100	44	29	31
Princeton, Mass.	.384	.395	.408	76	58	78	100	95	100	35	26	36
St. Louis, Mo.	.528	.526	.538	75	53	75	94	97	100	57	34	20
St. Martin, Ca., E.	.410	.525	.427	80	71	81	100	98	100	56	35	49
Savannah, Ga.	.690	.672	.688	79	56	78	92	92	95	57	31	52
Smithsonian Inst.	.551	.562	.597	83	60	81	100	100	100	59	32	48
Spencertown, N. Y. (3)	.420	.506	.444	82	75	84	100	100	100	37	48	37
Springdale, Ky.	.487	.709	.613	96	71	85	100	100	100	81	53	57
Upper Alton, Ill.	.568	.782	.575	89	81	84	100	100	94	76	31	59
Westfield, Mass.	.419	.451	.442	85	58	78	100	100	100	56	33	56
Worcester, Mass.	.430	.495	.423	73	65	74	100	100	100	33	39	43
Zanesville, Ohio.	.520	.565	.582	95	66	96	100	100	100	62	37	73

(1) Incomplete.
(2) Mean force of vapor, 775, computed approximately from the mean dew point.
(3) For the first 25 days of the month only.

NAME OF STATION.	Mean force of Vapor.			Relative Humidity.								
				Mean.			Maxima.			Minima.		
	7 A. M.	2 P. M.	9 P. M.	7 A.M.	2 P.M.	9 P.M.	7 A.M.	2 P.M.	9 P.M.	7 A.M.	2 P.M.	9 P.M.
Aiken, S. C.	.764	.735	.754	83	60	79	95	79	100	71	48	60
Alexandria, Va.	.760	.799	.780	86	68	86	100	91	100	58	55	53
All Saints, S. C.	.848	.868	.847	86	77	85	95	91	95	77	62	73
Amherst, Mass.	.646	.701	.647	94	73	91	100	99	100	84	48	62
Ann Arbor, Michigan (Winchell).	.578	.633	.589	83	67	89	97	95	99	65	41	72
Austin, Texas	.669	.559	.661	84	42	68	100	73	90	53	31	52
Bloomfield, N. J.	.688	.722	.692	87	66	89	98	97	97	72	42	68
Burlington, Vt.	.584	.625	.571	81	61	81	95	83	100	65	42	55
Camden, S. C.	.765	.742	.775	81	57	79	95	86	91	70	44	64
Charleston, S. C. (1).	.....	.....	.....	...	...	...	.....	.....	.....	...	...	...
Deaf & D'b Inst., N.Y.	.650	.671	.663	79	67	81	100	100	98	47	43	60
Dubuque, Iowa	.530	.562	.601	78	53	78	96	96	86	54	32	55
Frederick, Md.	.691	.704	.710	78	59	81	100	89	97	61	42	63
Gallipolis, Ohio	.715	.819	.783	84	73	85	91	95	100	76	56	76
Gardiner, Me.	.721	.765	.....	95	74	...	100	100	.....	81	49	...
Germantown, Ohio.	.687	.677	.674	87	63	87	96	98	99	71	36	71
Glenwood, Tenn.	.708	.766	.757	91	68	87	100	96	96	79	43	76
Green Springs, Ala. (2)	.786	.827	.817	85	63	85	91	87	100	74	41	68
Hillsboro', Ohio	.701	.753	.701	96	73	92	100	100	100	83	55	66
Jacksonville, Fla.	.901	.947	.884	89	78	90	100	100	100	79	60	79
Lima, Pa.	.601	.608	.601	86	66	90	100	96	98	60	46	77
Manchester, Ill.	.681	.924	.722	87	76	89	100	100	100	64	44	64
Muscatine, Iowa	.626	.759	.642	86	67	87	100	91	100	71	29	68
Nantucket, Mass.	.586	.589	.571	81	66	87	95	100	95	53	45	74
New Bedford, Mass.	.629	.677	.635	88	80	93	100	95	100	66	56	79
Newburyport, Mass.	.571	.585	.591	84	65	82	100	94	98	58	41	58
New Harmony, Ind.	.801	.834	.799	86	67	87	96	87	100	75	51	78
New Orleans, La.	.808	.797	.830	90	72	86	98	95	95	78	47	64
New Wied, Texas (2)	.836	.988	.858	85	62	92	100	92	100	56	39	67
Norristown, Pa.	.694	.688	.673	85	65	82	95	91	95	74	24	68
N'th Attleboro', Mass.	.603	.628	.571	81	65	84	96	94	96	56	42	17
Ottawa, Ill.	.728	.996	.712	94	88	94	100	100	100	78	79	88
Oxford, Miss.	.757	.774	.791	85	64	86	100	100	100	60	47	62
Penn Yan, N. Y.	.....	.573	.....	...	62	...	.....	95	.....	...	39	...
Philadelphia, Pa.	.714	.712	.740	80	62	74	94	89	94	55	43	57
Pomfret, Conn.	.655	.729	.649	95	82	95	100	100	100	85	65	77
Pottsville, Pa.	.578	.605	.615	71	52	86	100	84	100	43	27	47
Princeton, Mass.	.552	.569	.539	85	68	84	98	94	98	62	45	51
St. Louis, Mo.	.681	.683	.729	79	54	78	100	74	95	50	30	53
St. Martin, Can. East	.550	.637	.579	82	61	86	95	90	97	69	44	67
Savannah, Ga.	.804	.791	.809	82	57	80	94	84	99	70	42	67
Smithsonian Inst.	.717	.721	.763	83	60	82	100	93	100	63	39	61
Spencertown, N.Y. (2)	.676	.709	.707	90	73	92	100	83	100	86	45	86
Springdale, Ky.	.642	1.089	.851	97	86	94	100	100	100	81	47	72
Upper Alton, Ill.	.735	1.010	.752	94	86	91	100	100	100	85	59	75
Westfield, Mass.	.568	.587	.587	86	64	86	100	100	100	75	40	67
Worcester, Mass.	.527	.574	.527	71	62	70	100	100	100	45	32	45
Zanesville, Ohio	.700	.740	.735	96	64	95	100	100	100	80	37	67

* For changes in hours of observation, see tables of Thermometer for this month.
(1) Mean force of vapor, .907, computed approximately from the mean dew point.
(2) Month incomplete.

NAME OF STATION.	Mean force of Vapor.			Relative Humidity.								
				Mean.			Maxima.			Minima.		
	7 A. M.	2 P. M.	9 P. M.	7 A.M.	2 P.M.	9 P.M.	7 A. M.	2 P. M.	9 P. M.	7 A.M.	2 P.M.	9 P.M.
Aiken, S. C.	.763	.782	.761	85	65	80	95	90	100	78	53	62
Alexandria, Va.	.714	.755	.713	88	72	88	100	96	100	54	53	55
All Saints, S. C.	.824	.869	.834	88	76	84	95	91	95	59	51	56
Amherst, Mass.	.542	.543	.513	92	65	87	100	100	100	72	31	47
Ann Arbor, Michigan (Winchell).	.511	.574	.533	85	66	94	100	90	100	49	47	82
Auburn, Ala.	.688	.666	.668	80	60	76	95	100	100	66	42	61
Austin, Texas	.713	.589	.660	85	46	65	100	73	100	69	27	43
Bloomfield, N. J.	.544	.581	.555	86	60	85	98	82	98	62	32	54
Burlington, Vt.	.458	.546	.461	81	63	76	100	89	100	62	38	50
Camden, S. C.	.767	.768	.754	85	60	81	95	84	95	73	43	59
Charleston, S. C. (1)	.....	.....	.....	...	...	...	...	...	...	...	...	...
Deaf & D'b Inst., N. Y.	.537	.553	.557	73	62	76	95	95	95	46	37	49
Dubuque, Iowa.	.579	.663	.611	84	60	80	100	87	100	61	43	69
Frederick, Md.	.609	.670	.650	86	65	88	100	100	100	57	47	63
Gallipolis, Ohio.	.692	.775	.731	86	73	85	93	92	95	71	42	69
Germantown, Ohio.	.605	.680	.653	89	68	94	98	90	99	70	50	81
Gardiner, Maine (2).	.548	.580	.....	93	70	...	100	100	.....	70	37	...
Glenwood, Tenn. (2)	.731	.805	.770	92	77	90	100	97	96	83	51	75
Green Springs, Ala.	.794	.817	.821	87	68	83	100	91	95	65	48	56
Hillsboro', Ohio.	.612	.711	.657	95	76	96	100	100	100	89	52	84
Jacksonville, Fla.	.913	1.012	.931	91	80	91	100	96	100	70	66	83
Lima, Pa.	.577	.611	.578	86	67	88	98	91	98	58	40	61
Manchester, Ill.	.670	.796	.686	94	78	95	100	100	100	77	58	53
Muscatine, Iowa.	.615	.769	.606	86	69	88	100	91	100	66	49	68
Nantucket, Mass.	.505	.488	.487	70	59	78	97	92	100	35	25	37
New Bedford, Mass.	.517	.542	.527	86	68	86	100	95	100	62	45	36
Newburyport, Mass.	.451	.485	.493	81	55	79	95	91	98	59	25	57
New Harmony, Ind.	.702	.809	.758	89	72	92	100	95	100	70	54	85
New Orleans, La.	.839	.807	.853	93	67	84	100	98	98	75	46	66
New Wied, Texas.	.886	1.047	.915	89	71	94	100	100	100	74	44	68
Norristown, Pa.	.548	.609	.571	82	67	82	95	85	95	57	49	56
N'th Attleboro', Mass.	.461	.484	.470	80	55	86	95	98	98	57	30	64
Ottawa, Ill.	.607	.915	.637	93	90	94	100	96	100	70	82	87
Oxford, Miss. (3).	.853	.835	.842	91	68	90	100	100	100	78	43	79
Penn Yan, N. Y.	.....	.498	.....	...	59	...	.....	85	.....	...	36	...
Philadelphia, Pa.	.603	.616	.619	75	62	72	91	86	94	55	39	49
Pomfret, Conn.	.519	.612	.525	91	75	92	100	100	100	75	45	75
Port Gibson, Miss. (3)	.831	.869	.828	87	71	77	95	87	91	77	45	47
Pottsville, Pa.	.455	.503	.521	70	53	84	100	86	100	29	33	55
Princeton, Mass.	.441	.448	.415	80	58	75	96	95	95	60	33	45
St. Louis, Mo.	.642	.682	.685	86	65	85	97	92	100	69	38	65
St. Martin, Can. East	.413	.552	.447	85	59	79	99	96	93	64	36	58
Savannah, Ga.	.814	.811	.820	84	61	80	95	89	94	76	42	67
Smithsonian Inst.	.649	.945	.666	87	59	80	95	82	100	56	39	46
Spencertown, N. Y.	.493	.586	.518	84	71	87	100	95	100	69	48	50
Springdale, Ky.	.603	.878	.760	97	76	87	100	100	100	94	38	69
Upper Alton, Ill.	.619	.862	.661	91	88	92	100	100	100	50	66	74
Westfield, Mass.	.448	.517	.491	87	59	85	100	100	100	56	38	69
Worcester, Mass.	.420	.566	.409	71	66	66	100	100	86	47	39	48
Zanesville, Ohio.	.572	.613	.665	92	58	91	100	90	100	30	29	56

* For changes in hours of observation, see tables of Thermometer for this month.
(1) Mean force of vapor, .911, computed approximately from the mean dew point.
(2) Month incomplete.
(3) Record incomplete.

NAME OF STATION.	Mean force of Vapor.			Relative Humidity.								
				Mean.			Maxima.			Minima.		
	7 A. M.	2 P. M.	9 P. M.	7 A.M.	2 P.M.	9 P.M.	7 A. M.	2 P. M.	9 P. M.	7 A.M.	2 P.M.	9 P.M.
Aiken, S. C.	.696	.675	.642	86	59	70	100	95	90	72	47	52
Alexandria, Va.	.593	.676	.627	91	75	89	100	100	100	76	63	77
All Saints, S. C.	.733	.770	.742	89	73	81	95	83	96	69	56	63
Amherst, Mass.	.417	.463	.451	95	63	82	100	89	100	76	45	70
Ann Arbor, Michigan, (Winchell)	.491	.560	.487	91	72	91	100	96	100	73	33	69
Auburn, Ala.	.668	.611	.643	76	52	72	100	100	100	63	29	47
Augusta, Ga.	.763	.808	.756	80	67	75	100	87	100	61	51	44
Austin, Texas.	.669	.638	.686	90	52	74	100	78	90	49	37	60
Bloomfield, N. J.	.442	.526	.477	86	59	86	100	90	100	64	34	58
Burlington, Vt.	.366	.455	.389	80	62	76	100	83	90	53	43	55
Camden, S. C.	.663	.694	.695	84	58	79	95	81	95	68	44	62
Charleston, S. C. (1).	......	......	......	....	....	....	.....	.....	.....	....	....	....
Deaf & D'b Inst., N.Y	.442	.484	.475	71	58	74	92	85	96	54	42	54
Dubuque, Iowa.	.510	.522	.514	87	66	81	100	100	100	68	40	63
Frederick, Md.	.538	.650	.609	93	75	92	100	100	100	80	62	79
Gallipolis, Ohio.	.631	.736	.672	88	75	85	94	95	95	81	53	62
Germantown, Ohio.	.598	.652	.591	93	74	93	98	95	99	73	44	57
Glenwood, Tenn.	.692	.709	.723	93	69	92	99	98	100	83	40	79
Hillsboro', Ohio.	.606	.682	.632	99	82	98	100	100	100	95	61	81
Hiram, Ohio.	.498	.589	.512	79	77	76	88	95	88	71	61	54
Jacksonville, Fla.	.902	.966	.891	92	81	91	96	95	96	87	65	83
Lima, Pa.	.476	.527	.509	90	66	88	100	93	97	73	38	73
Manchester, Ill.	.607	.758	.639	97	81	94	100	100	100	80	59	80
Montreal, Canada.	.359	.381	.377	75	58	71	94	95	94	46	22	32
Muscatine, Iowa.	.604	.731	.600	92	79	91	100	100	100	79	54	52
Nantucket, Mass.	.407	.400	.420	69	55	77	96	97	91	32	12	40
New Bedford, Mass.	.421	.485	.430	88	67	87	100	95	100	61	38	63
Newburyport, Mass.	.377	.400	.391	81	56	77	100	98	100	55	30	49
New Harmony, Ind.	.695	.796	.736	93	76	91	100	95	95	79	65	70
New Orleans, La.	.830	.805	.849	93	73	89	100	100	100	77	39	67
New Wied, Texas (2)	.829	1.052	.858	90	75	94	100	100	100	77	58	87
Norristown, Pa.	.442	.506	.464	84	64	78	94	90	100	52	38	44
N'th Attleboro', Mass.	.368	.371	.382	77	46	79	95	71	94	56	23	60
Ottawa, Ill.	.602	.770	.597	97	87	92	100	100	100	73	2	74
Oxford, Miss.	.752	.797	.767	87	75	93	100	100	100	75	49	55
Penn Yan, N. Y.	......	.438	......	....	60	....	.....	93	.....	....	31	....
Philadelphia, Pa.	.515	.551	.563	79	61	76	94	89	94	63	45	54
Pomfret, Conn.	.421	.512	.435	90	73	90	100	100	100	70	58	72
Pottsville, Pa.	.402	.512	.520	76	59	92	97	100	100	41	32	72
Princeton, Mass.	.381	.376	.355	79	57	78	97	95	94	31	32	54
St. Louis, Mo.	.639	.674	.654	89	65	84	100	93	98	75	42	69
St. Martin, Can. East	.357	.451	.383	88	62	81	99	92	98	55	35	50
Savannah, Ga.	.773	.762	.799	89	63	84	95	97	92	77	40	73
Smithsonian Inst.	.548	.574	.621	81	61	86	100	100	100	50	44	60
Spencertown, N. Y.	.377	.450	.427	84	63	86	100	89	100	65	39	73
Springdale, Ky.	.610	.710	.714	97	72	88	100	95	100	88	54	62
Upper Alton, Ill.	.624	.867	.620	94	89	89	100	100	100	85	74	53
Westfield, Mass.	.378	.410	.440	87	54	84	100	91	95	66	22	70
Worcester, Mass.	.413	.502	.367	71	63	70	94	100	100	44	11	54
Zanesville, Ohio.	.541	.579	.557	85	64	80	100	86	100	39	41	44

* For changes in hours of observation, see tables of Thermometer for this month.

(1) Mean force of vapor, .851, computed approximately from the mean dew point.

(2) Month incomplete.

NAME OF STATION.	Mean force of Vapor.			Relative Humidity.								
				Mean.			Maxima.			Minima.		
	7 A. M.	2 P. M.	9 P. M.	7 A.M.	2 P.M.	9 P.M.	7 A. M.	2 P. M.	9 P. M.	7 A.M.	2 P.M.	9 P.M.
Aiken, S. C.	.362	.360	.371	84	46	66	100	87	100	57	14	42
Alexandria, Va.	.312	.402	.325	94	76	87	100	100	100	74	57	63
All Saints, S. C.	.419	.496	.474	83	64	78	95	95	94	46	39	45
Amherst, Mass.	.317	.348	.327	96	74	89	100	100	100	73	42	56
Ann Arbor, Mich.	.233	.267	.248	83	64	87	100	98	100	62	36	42
Auburn, Ala.	.296	.298	.327	64	39	60	88	78	100	10	12	34
Austin, Texas.	.416	.453	.464	90	50	83	100	75	100	68	12	15
Bloomfield, N. J.	.309	.318	.308	86	63	82	97	97	97	63	17	56
Burlington, Vt.	.260	.273	.258	79	60	72	100	90	100	53	32	19
Camden, S. C.	.324	.395	.364	82	51	79	94	100	89	57	27	52
Camden, Ark. (1)	.399	.479	.422	81	65	78	91	79	89	65	42	67
Deaf & D'bInst., N.Y.	.293	.335	.308	74	66	74	100	91	100	38	41	49
Dubuque, Iowa.	.214	.205	.214	79	48	66	100	97	100	44	17	14
Frederick, Md.	.278	.330	.286	87	62	78	100	100	100	49	39	53
Gallipolis, Ohio.	.291	.344	.309	88	61	81	100	92	100	71	17	51
Germantown, Ohio.	.243	.289	.272	91	57	88	98	90	98	67	41	41
Glenwood, Tenn.	.309	.330	.333	88	51	84	97	96	98	60	26	38
Hillsboro', Ohio.	.281	.326	.308	96	67	90	100	100	100	72	27	47
Hiram, Ohio.	.294	.371	.301	68	68	62	83	83	76	45	44	39
Jacksonville, Fla.	.516	.666	.560	89	78	91	100	100	100	59	65	77
Lima, Pa.	.289	.322	.307	89	65	85	100	100	97	50	34	63
Manchester, Ill.	.269	.396	.299	92	70	88	100	100	100	62	40	62
Montreal, Canada.	.270	.298	.301	82	71	87	100	100	100	41	40	60
Muscatine, Iowa.	.201	.301	.229	79	59	76	100	92	100	53	31	34
Nantucket, Mass.	.358	.361	.363	78	66	83	100	100	98	47	26	51
New Bedford, Mass.	.333	.383	.354	84	73	86	100	100	100	50	41	61
Newburyport, Mass.	.301	.333	.313	88	72	85	100	100	100	55	26	67
New Harmony, Ind.	.293	.378	.336	95	65	88	100	100	100	65	39	51
New Orleans, La.	.462	.460	.482	85	63	73	100	100	100	39	34	45
New Wied, Texas.	.467	.652	.558	86	68	86	100	100	100	56	33	42
Norristown, Pa.	.279	.333	.299	84	65	79	100	90	94	62	33	61
N'th Attleboro', Mass	.282	.299	.303	81	54	80	97	98	100	59	23	31
Ottawa, Ill., (2)	.271	.524	.323	99	89	95	100	100	100	90	77	63
Oxford, Miss.	.380	.523	.411	96	78	90	100	100	100	83	60	53
Philadelpha, Pa.	.324	.348	.342	82	61	78	97	97	97	56	42	60
Pomfret, Conn.	.306	.372	.322	93	79	89	100	100	100	73	51	53
Pottsville, Pa.	.242	.296	.267	78	66	89	100	100	100	12	32	58
Princeton, Mass.	.270	.282	.275	84	67	79	100	100	100	62	44	42
St. Louis, Mo.	.253	.283	.284	78	49	70	100	87	96	41	24	31
St. Martin, Can. East	.247	.325	.284	93	75	89	100	98	100	69	38	66
Savannah, Ga.	.413	.420	.441	81	48	75	94	84	88	64	26	49
Smithsonian Inst.	.308	.340	.342	87	61	84	100	100	100	51	29	54
Spencertown, N. Y.	.265	.305	.282	87	71	83	100	100	94	70	32	68
Springdale, Ky.	.255	.366	.361	94	63	76	100	100	100	70	38	29
Upper Alton, Ill.	.246	.421	.319	83	77	84	94	94	94	65	59	51
Warrington, Fla.	.658	.759	.712	93	88	89	100	100	100	73	59	75
Westfield, Mass.	.272	.305	.309	86	66	87	100	100	100	63	36	59
Woodward H.School, Ohio.	.335	.385	.382	82	66	80	100	100	100	67	2	52
Zanesville, Ohio.	.228	.310	.271	79	63	77	100	100	100	31	26	30

(1) For the first 12 days of the month only.
(2) For the first 21 days of the month only.

NAME OF STATION.	Mean force of Vapor.			Relative Humidity.								
				Mean.			Maxima.			Minima.		
	7 A. M.	2 P. M.	9 P. M.	7 A.M.	2 P.M	9 P.M.	7 A. M.	2 P.M.	9 P.M.	7 A.M.	2 P.M.	9 P.M.
Aiken, S. C..........	.400	.417	.431	89	59	82	100	100	100	66	30	58
Alexandria, Va.......	.300	.344	.303	97	79	89	100	100	100	72	55	72
All Saints, S. C.......	.398	.468	438	89	68	85	100	92	100	67	36	57
Amherst, Mass.......	.192	.225	.209	87	70	85	100	100	100	53	43	52
Ann Arbor, Mich......	.194	.216	.201	88	71	83	100	100	100	72	39	68
Auburn, Ala..........	.353	.376	.372	74	59	67	94	90	94	43	24	32
Austin, Texas.........	.356	.310	.375	90	42	83	100	100	100	62	....	56
Blackwell's Isla'd, N.Y.	.201	.210	.203	69	57	62	93	93	93	35	....	19
Bloomfield, N. J. (1)...	.219	.253	.237	78	65	75	96	97	97	60	39	37
Burlington, Vt.......	.153	.177	.158	73	65	71	100	100	89	49	37	40
Camden, Ark..........	.320	.392	.358	80	60	74	100	86	95	61	34	7
Camden, S. C.........	.339	.395	.365	85	60	77	94	94	93	66	18	45
Deaf & Dumb Inst., N.Y	.215	.242	.220	75	67	70	100	97	97	47	41	47
Detroit, Mich. (2).....	.203	.245	.220	86	81	83	100	100	100	68	56	46
Dubuque, Iowa (1)....	.174	.185	.191	74	60	68	95	100	89	52	28	17
Frederick, Md.........	.238	.260	.249	82	64	70	100	100	98	50	34	50
Gallipolis, Ohio.......	.267	.333	.276	84	67	78	100	94	93	61	29	57
Germantown, Ohio....	.241	.264	.247	89	69	85	100	100	100	71	34	52
Glenwood, Tenn.......	.310	.313	.303	82	58	76	100	100	99	42	27	48
Hillsborough, Ohio...	.258	.301	.259	92	71	85	100	100	100	61	36	49
Lima, Pa............	.233	.256	.239	83	64	83	98	98	100	51	33	52
Manchester, Ill.......	.236	.309	.236	96	80	92	100	100	100	69	44	63
Montreal, Can........	.168	.185	.173	87	78	85	100	100	100	62	55	7
Muscatine, Iowa......	.158	.236	.162	70	76	69	100	100	100	....	42	5
Nantucket, Mass......	.251	.244	.262	76	66	81	96	96	100	50	38	55
New Bedford, Mass...	.226	.244	.239	84	71	84	100	100	100	50	37	60
Newburybort, Mass...	.187	.200	.202	83	65	84	100	100	100	52	30	51
New Harmony, Ind...	.286	.291	.308	92	65	88	100	94	100	66	41	51
New Orleans, La......	.514	.542	.541	89	76	86	100	100	100	59	35	55
New Wied, Texas	.435	.506	.448	91	58	86	100	90	100	74	29	54
Norristown, Pa.......	.214	.253	.219	74	66	71	100	88	100	39	31	33
North Attleboro', Mass.	.184	.205	.200	77	60	76	100	100	100	32	25	37
Oxford, Miss..........	.362	.464	.389	89	79	85	100	100	100	71	7	28
Perrysburg, Ohio.....	.228	.272	.237	84	73	81	100	93	100	61	50	65
Philadelphia, Pa......	.247	.266	.253	79	60	73	94	97	97	56	28	44
Pomfret, Conn.........	.204	.261	.222	88	81	88	100	100	100	45	53	70
Pottsville, Pa........	.224	.228	.221	82	66	84	96	95	100	47	30	58
Princeton, Mass.......	.169	.173	.164	81	70	74	100	100	100	45	30	31
St. Louis, Mo.........	.226	.236	.242	80	57	72	100	96	96	63	26	45
Savannah, Ga........	.442	.493	.435	85	65	82	97	93	94	60	57	55
Smithsonian Inst......	.267	.282	.291	87	64	82	100	100	100	48	25	38
Spencertown, N. Y. (3)	.223	.244	.231	91	80	82	100	100	94	61	39	59
Springdale, Ky.......	.279	.321	.309	92	71	78	100	100	100	71	38	54
Upper Alton, Ill......	.296	.382	.324	88	80	86	100	94	100	69	59	66
Westfield, Mass......	.189	.201	.212	84	64	86	100	100	100	58	19	45
Woodw'd H. School, O.	.289	.326	.287	89	75	85	100	100	100	65	44	59
Zanesville, Ohio......	.217	.251	.232	77	65	76	100	83	100	58	39	43

(1) Incomplete.
(2) For the last 20 days of the month only.
(3) For the first 19 days of the month only.

NAME OF STATION.	Mean force of Vapor.			Relative Humidity.								
				Mean.			Maxima.			Minima.		
	7 A. M.	2 P. M.	9 P. M.	7 A.M.	2 P.M.	9 P.M.	7 A. M.	2 P.M.	9 P.M.	7 A.M.	2 P.M.	9 P.M.
Alexandria, Va.......	.199	.237	.194	93	82	82	100	100	100	65	55	42
All Saints, S. C.......	.305	.366	.342	89	75	88	100	100	100	61	32	51
Amherst, Mass........	.130	.144	.143	90	74	87	100	100	100	60	41	53
Ann Arbor, Mich.....	.121	.141	.122	91	85	88	100	100	100	67	59	71
Austin, Texas........	.228	.265	.304	88	54	83	100	100	100	53	16	40
Bloomfield, N. J. (1)..	.145	.160	.187	76	58	79	95	94	96	50	25	50
Burlington, Vt.......	.091	.110	.093	61	60	56	100	82	82	....	23	....
Camden, S. C.........	.239	.290	267	87	66	83	100	100	94	57	34	48
Deaf & Dumb Inst., N.Y	.117	.147	.133	58	59	56	94	97	92	16	26	9
Detroit, Mich........	.145	.155	.146	93	83	90	100	100	100	61	46	61
Frederick, Md.	.142	.165	.150	79	62	74	100	97	100	47	37	....
Germantown, Ohio....	.136	.142	.137	89	68	82	100	96	98	60	31	46
Glenwood, Tenn......	.168	.166	.180	80	57	75	100	95	95	43	14	29
Hillsborough, Ohio...	.162	.171	.160	94	78	88	100	100	100	73	39	49
Jacksonville, Fla......	.406	.551	.449	91	86	90	95	95	100	78	70	47
Lima, Pa............	.144	.171	.160	85	64	82	100	100	100	44	33	49
Manchester, Ill.......	.130	.183	.143	98	85	95	100	100	100	78	51	67
Montreal, Can........	.116	.131	.129	81	76	82	100	100	100	55	54	55
Nantucket, Mass......	.186	.152	.193	77	66	77	100	93	93	40	37	50
Nazareth Hall, Pa....	.138	.167	.150	98	85	95	100	100	100	77	9	70
New Bedford, Mass...	.156	.153	.175	84	64	82	100	100	100	58	36	46
Newburyport, Mass. (2)	.128	.138	.140	84	68	87	100	100	100	....	36	63
New Harmony, Ind...	.163	.183	.170	88	71	86	100	100	100	44	42	51
New Wied, Texas.....	.258	.391	.322	73	65	79	100	93	100	....	19	6
Norristown, Pa.......	.116	.130	.130	73	58	70	100	100	100	49	15	38
North Attleboro', Mass.	.124	.119	.148	75	55	77	100	92	94	46	25	51
Oxford, Miss.........	.213	.283	.228	91	80	87	100	100	100	70	42	50
Perrysburg, Ohio.....	.139	.152	.148	85	76	85	100	100	100	59	39	46
Philadelphia, Pa......	.161	.182	.176	78	66	74	97	92	100	52	40	42
Pomfret, Conn. (2).,..	.150	.175	.168	90	77	85	100	100	100	66	46	62
Pottsville, Pa........	.123	.146	.129	83	69	81	100	100	100	40	34	20
Princeton, Mass......	.121	.113	.122	84	70	78	100	100	100	49	30	47
St. Louis, Mo.........	.142	.138	.143	82	58	74	100	93	96	51	16	40
St. Martin's, Can......	.099	.122	.113	90	82	93	99	100	99	....	28	75
Savannah, Ga........	.314	.368	.351	83	69	82	98	96	96	56	22	28
Smithsonian Inst......	.166	.190	.185	83	66	81	100	100	100	45	31	37
Springdale, Ky.......	.162	.208	.192	93	81	88	100	100	100	62	43	65
Westfield, Mass... ...	.113	.122	.142	82	61	88	100	100	100	55	25	52
Woodw'd H. School, O.	.179	.222	.197	80	75	76	100	100	100	52	41	29
Zanesville, Ohio (3)...	.160	.152	.152	87	53	72	100	67	100	53	39	48

(1) For the first 24 days of the month only.
(2) Incomplete.
(3) For the first 8 days of the month only.

NAME OF STATION.	7 A. M.	2 P. M.	9 P. M.	NAME OF STATION.	7 A. M.	2 P. M.	9 P. M.
Alexandria, Va	7.2	6.4	4.4	Jackson, Ohio	7.6	7.9	6.0
All Saints, S. C	5.2	4.4	3.2	Jacksonville, Fa	5.1	4.4	3.3
Amherst, Mass	6.6	6.5	6.1	Janesville Wis.	7.2	6.7	6.7
Angelica, N. Y	6.5	6.0	6.5	Keokuck, Iowa	5.4	5.6	5.4
Ann Arbor, Mich	8.9	7.7	7.7	Key West, Fa	3.2	3.4	....
Ann Arbor, " (Winchell)	8.3	7.0	7.4	Knox Hill, Fa.	4.6	3.8	2.3
Arcola, Ohio	8.5	6.9	7.7	Knoxville, Tenn	3.3	6.2	4.9
Ashland, Va	6.2	5.9	5.2	La Grange, Ga (2)	4.8	6.6	6.4
Athens, Ill	4.8	5.7	6.2	Lewisburg, Va	6.4	6.5	7.1
Auburn, Ala	5.0	4.3	2.4	Lima, Pa	6.8	6.5	5.3
Augusta, Ill	5.6	5.3	5.6	Lodi, N. Y	7.8	7.9	7.6
Austin, Texas	5.8	5.2	3.1	Londonderry, N. H.	6.2	6.3	5.9
Baldwinsville, N. Y	8.4	8.5	6.6	Lowville, N. Y	7.0	6.8	7.2
Ballardsville, Ky	7.2	7.5	4.9	Madison, Ia.	7.2	7.0	6.0
Battle Creek, Mich	8.0	7.4	8.0	Manchester, N. H	6.5	6.7	6.2
Bedford, Pa.	7.0	5.8	5.5	Manchester, Ill	5.5	6.1	6.4
Beloit, Wis.	4.3	4.6	4.9	Marietta, Ohio	5.9	5.9	6.1
Beverly, N. Y	7.0	7.2	5.3	Mendon, Mass	6.7	7.1	5.7
Bladensburg, Md	8.0	6.2	4.9	Millersburg, Ky	6.4	6.3	6.3
Bloomfield, N. J.	6.6	6.2	6.0	Milton, Ia	6.2	5.5	5.3
Brandon, Vt	6.9	6.9	6.5	Milwaukie, Wis	5.6	5.0	5.6
Burlington, Vt	4.1	3.9	3.2	Monroe, (Whelply) Mich.	7.4	7.5	6.6
Burlington, N. J	8.9	6.7	7.7	Montcalm, Va	8.3	7.4	8.3
Camden, S. C	6.1	5.3	4.8	Morrisville, Pa	6.3	5.2	4.3
Canton, N. Y	7.5	7.8	7.0	Moss Grove, Pa	6.7	6.4	5.8
Carmel, Me	5.7	5.9	4.2	Mt. Vernon, Ohio	6.6	6.8	6.1
Castleton, Vt	8.1	8.4	7.7	Muscatine, Iowa	3.4	3.9	4.0
Cedar Keys, Fa	4.3	4.1	3.0	Nantucket, Mass.	6.6	6.8	5.8
Ceresco, Wis	8.7	7.4	8.2	Newark, Ohio	7.1	8.0	5.8
Chapel Hill, N. C	4.4	4.5	6.3	New Bedford, Mass.	7.8	7.1	7.6
Cincinnati, Ohio. (1)	....	....	....	Newburyport, Mass.	8.4	7.8	8.7
Concord, Mass	5.6	6.2	5.6	New Harmony, Ind	6.0	6.6	6.1
Cooper, Mich	7.8	6.8	7.0	New Lisbon, Ohio	6.4	6.4	5.7
Crichton's Store, Va	5.7	5.4	5.9	New Orleans, La	5.3	5.7	2.5
Danville, Ky	6.2	5.2	5.2	New Wied, Texas	4.5	5.4	6.1
Deaf and Dumb Inst., N. Y.	7.2	6.3	5.0	Norristown, Pa	7.0	6.2	4.7
Detroit, Mich	8.7	8.0	7.7	North Attleboro, Mass.	6.1	6.5	6.0
Dubuque, Iowa	6.3	6.7	6.7	Norwalk, Ohio	6.0	6.8	5.2
Easton, Pa.	6.9	6.8	5.2	Oberlin, Ohio	6.3	5.8	5.7
Exeter, N. H	6.1	5.6	6.1	Oldtown, Me.	3.4	3.2	2.0
Fort Madison, Iowa	5.1	5.0	5.3	Ottawa, Ill	1.4	2.3	3.1
Frederick, Md	7.2	6.1	5.5	Oxford, Miss	5.4	6.1	3.8
Fryeburg, Me	5.8	5.8	5.4	Penn Yan, N. Y	8.2	7.8	7.2
Gardiner, Me	7.0	7.5	6.1	Perry, Me.	5.4	5.5	6.3
Garlandsville, Miss	4.5	3.7	3.0	Philadelphia, Pa	7.3	6.7	5.0
Germantown, Ohio	7.7	7.8	6.5	Platteville, Wis	6.0	6.5	5.9
Gettysburg, Pa	6.8	5.7	6.5	Pocopson, Pa	7.4	6.9	5.2
Glenwood, Tenn.	7.0	7.1	4.1	Pomfret, Conn.	6.2	6.4	5.6
Gouverneur, N. Y	8.4	8.7	8.2	Pottsville, Pa	7.4	6.4	6.0
Grand Rapids, Mich	7.1	7.0	6.0	Poultney, Iowa	6.3	6.5	6.5
Granville, Ohio	6.3	7.1	66.	Princeton, Mass.	7.0	7.0	6.3
Green Springs, Ala	5.1	4.3	2.9	Providence, R. I	6.0	8.4	6.4
Great Falls, N. H	6.0	7.0	6.3	Quasqueton, Iowa	6.0	5.9	6.0
Hannibal, Mo	6.2	6.0	6.2	Randolph, Pa	8.0	6.5	7.0
Harrisburg, Pa	7.5	6.1	6.2	Richmond, Mass	8.0	7.3	8.1
Hillsboro, Ohio.	7.0	5.7	....	Richmond, Ohio	7.3	6.5	6.0

* For changes in hours of observation see tables of Thermometer for this month.

(1) Mean for the month 1.5.

(2) Obs. at sunrise, noon and sunset.

NAME OF STATION.	7 A. M.	2 P. M.	9 P. M.	NAME OF STATION.	7 A. M.	2 P. M.	9 P. M.
Sag Harbor, N. Y.........	6.4	6.3	5.2	Springfield, Mass.	6.0	6.5	5.9
St. James, Mich............	8.7	8.4	8.3	Steuben, Me.	5.9	7.2	8.1
St. Johnsbury, Vt.........	4.7	4.9	5.4	Thornbury, N. C...........	6.0	4.8	3.9
St. Joseph's, Min..........	6.2	5.9	5.0	Tuscaloosa, Ala	4.5	4.0	3.4
St. Louis, Mo..............	6.0	5.5	5.5	Upper Alton, Ill............	1.8*	2.7	0.7
St. Martin, Ca., E.........	6.5	5.2	5.4	Wampsville, N. Y..........	8.0	7.9	7.3
Salmon Falls, N. H........	6.9	6.4	5.9	Warrington Fa.............	2.9	2.9	2.6
Savannah, Ga..............	5.6	5.0	3.3	West Haverford, Pa........	5.7	6.3	....
Saybrook, Ct...............	3.8	6.5	4.3	Whitemarsh Isl'd, Ga.....	4.4	4.0	3.6
Schellman Hall, Md........	5.9	5.8	5.0	Williamstown, Mass........	7.7	6.5	7.6
Smithville, N. Y...........	7.2	7.3	6.3	Winchester, Va.............	5.2	4.4	4.5
Southwick, Mass...........	7.1	7.7	8.0	Woodward, H'h School, O.	6.8	6.2	4.8
Sparta, Ga..................	5.2	6.4	4.0	Worcester, Mass............	6.4	6.6	5.3
Spencertown N. Y.........	6.6	6.3	6.4	Zanesville, Ohio...........	7.3	7.1	6.8
Springdale, Ky.............	7.1	7.5	6.9	....	....	....	....

NAME OF STATION.	7 A. M.	2 P. M.	9 P. M.	NAME OF STATION.	7 A. M.	2 P. M.	9 P. M.
Alexandria, Va	5.2	5.3	4.8	Key West, Fa	4.9	4.1	....
All Saints, S. C	3.7	3.4	2.5	Knoxville, Tenn	5.7	6.0	5.1
Amherst, Mass	5.7	5.5	4.7	Knox Hill, Fa	4.9	4.6	2.9
Angelica, N. Y	8.2	6.8	6.8	Laquiparle, Min	4.4	4.6	4.3
Ann Arbor, Mich., (A.W.)	8.1	6.6	6.5	Lewisburg, Va	7.9	6.8	5.4
Ann Arbor, (Woodruff)	8.3	6.5	3.5	Lima, Pa	5.3	6.2	6.1
Athens, Ill	4.9	5.4	4.7	Lodi, N. Y	7.5	9.1	7.6
Augusta, Ill	5.7	4.7	4.8	Londonderry, N. H	4.8	4.8	5.5
Austin, Texas	5.8	5.2	3.5	Lowville, N. Y	6.9	6.8	6.9
Auburn, Ala	4.0	3.2	2.3	Madison, Ohio	8.5	7.9	6.7
Baldwinsville, N. Y	7.5	8.1	7.9	Madison, Ind	8.0	7.3	6.7
Ballardsville, Ky	6.9	7.4	5.6	Manchester, N. H	5.9	5.5	6.1
Battle Creek, Mich	7.3	7.4	6.5	Manchester, Ill.	5.6	6.2	5.0
Bedford, Pa	7.7	7.1	7.7	Marietta, Ohio	7.5	7.1	6.3
Beloit, Wis	3.8	3.7	2.6	Mendon, Mass	4.9	6.1	5.5
Beverly, N. Y	6.0	6.3	5.4	Millersburg, Ky	6.5	7.2	5.1
Bladensburg, Md	6.0	3.8	5.5	Milton, Ind	6.8	5.8	4.7
Bloomfield, N. J	4.1	4.9	5.3	Milwaukie, Wis. (Winkler)	4.5	4.4	4.5
Boston, Mass.(1)	....	....	....	Monroe, Mich. (H. W.)	8.3	7.9	7.5
Brandon, Vt	5.9	5.5	6.1	Mount Vernon, Ohio	8.1	7.2	4.3
Burlington, Vt	3.6	3.2	3.7	Morrisville, Pa	3.6	3.6	4.4
Burlington, N. J	8.0	7.0	8.0	Muscatine, Iowa.	2.4	2.0	2.3
Camden, S. C	4.6	4.6	4.9	Nantucket, Mass	6.5	5.9	5.7
Canton, N. Y	7.1	7.7	8.1	Newark, Ohio	8.7	8.0	4.7
Carmel, Me	5.2	6.0	5.5	New Bedford, Mass	5.8	5.8	4.5
Castleton, Vt	7.6	6.7	7.0	Newburyport, Mass	8.7	8.0	9.9
Cedar Keys, Fa	3.1	3.6	3.1	New Harmony, Ind	4.6	5.7	4.6
Ceresco, Wis	6.0	6.3	5.3	New Lisbon, Ohio	7.9	8.1	6.9
Chapel Hill, N. C	5.7	5.2	6.9	New Orleans, La	5.2	4.9	2.5
Cincinnati, Ohio(2)	....	....	....	New Wied, Texas	5.0	5.6	6.1
Concord, N. H	4.9	4.6	5.3	Norristown, Pa	5.9	6.9	6.4
Cooper, Mich	8.6	8.6	6.8	North Attleboro, Mass	4.5	5.0	4.5
Danville, Ky	5.5	4.7	4.8	Oberlin, Ohio	7.8	7.2	6.7
Detroit, Mich	8.2	7.9	6.8	Oldtown, Me	2.6	1.9	1.3
Dubuque, Iowa	5.6	4.5	5.0	Ottawa, Ill	3.0	2.2	2.9
Easton, Pa	5.6	6.1	5.5	Oxford, Miss	3.7	3.4	2.6
Exeter, N. H	5.7	5.1	4.8	Penn Yan, N. Y	7.5	8.6	7.2
Fort Madison, Iowa	6.0	4.5	4.5	Perry, Me	6.0	5.5	4.1
Frederick, Md	6.1	5.8	6.0	Philadelphia, Pa.	5.4	6.1	6.1
Fryeburg, Me	5.4	4.1	5.0	Platteville, Wis	4.7	4.5	4.2
Gardiner, Me	6.5	6.4	6.5	Pocopson, Pa	6.6	6.6	6.8
Garlandsville, Miss	5.2	4.5	2.9	Pomfret, Conn	5.0	4.7	4.4
Germantown, Ohio.	7.8	7.6	5.6	Pottsville, Pa.	6.9	6.5	6.2
Gettysburg, Pa	5.9	5.6	6.6	Poultney, Iowa	5.4	5.7	5.3
Glenwood, Tenn	5.6	6.3	4.2	Princeton, Mass	5.5	5.9	4.7
Gouverneur, N. Y	9.2	9.0	8.1	Providence, R. I	4.6	4.7	4.0
Grand Rapids, Mich	7.3	6.5	5.0	Quasqueton, Iowa	6.1	5.2	4.5
Granville, Ohio	8.4	8.4	5.2	Randolph, Pa	8.3	8.5	8.2
Great Falls, N. H	5.9	5.5	4.7	Richmond, Ohio	8.1	7.6	6.7
Green Springs, Ala	4.0	3.2	2.4	Sag Harbor, N. Y	5.0	5.1	4.5
Hannibal, Me	5.4	4.8	3.3	St. James, Mich	6.3	5.4	5.2
Harrisburg, Pa	5.9	6.5	6.5	St. Louis, Mo	5.1	4.6	4.2
Hillsboro, Ohio	7.0	5.9	....	St. Joseph's, Min	5.7	6.2	4.6
Jackson, Ohio	8.0	7.2	6.0	St. Martin, Ca. E	5.7	5.1	5.1
Jacksonville, Fa	3.0	3.0	1.6	Savannah, Ga	4.8	4.1	3.5
Janesville, Wis	5.1	4.4	5.2	Schellman Hall, Md	5.5	5.1	5.5

* For changes in hours of observation see notes for tables of Thermometer for this month.
(1) Monthly mean, 4.2.
(2) Monthly mean, 5.7.

NAME OF STATION.	7 P. M.	2 P. M.	9 P. M.	NAME OF STATION.	7 A. M.	2 P. M.	9 P. M.
Smithville, N. Y.	6.5	8.8	6.8	Wampsville, N. Y.	7.9	8.4	8.7
Southwick, Mass.	6.6	7.5	9.2	Warrington, Fa.	3.6	2.9	2.8
Sparta, Ga.	5.5	5.5	4.0	West Haverford, Pa.	5.2	6.1	....
Spencertown, N. Y.	5.8	5.9	5.7	West Stockbridge, Mass.	6.7	7.1	5.5
Springdale, Ky.	6.0	6.9	6.0	Whitemarsh Isl'd, Ga.	3.7	3.9	3.3
Springfield, Mass.	5.5	5.7	4.4	Williamstown, Mass.	8.3	7.2	6.1
Steuben, Me	7.2	7.4	7.0	Winchester, Va.	5.4	4.8	4.0
Taunton, Mass.	2·1	4.0	2.8	Woodward's High School.	7.8	6.2	5.1
Thornbury, N. C.	4.6	4.8	2.5	Wood's Hole, Mass.	5.6	5.5	5.3
Tuscaloosa, Ala.	3.4	2.7	3.6	Worcester, Mass.	4.1	4.7	3.3
Upper Alton, Ill.	3.4	4.2	2.0	Zanesville, Ohio	8.7	8.2	4.1
Urbana, Ohio	8.3	7.5	6.1	....	....	....	....

NAME OF STATION.	7 A. M.	2 P. M.	9 P. M.
Alexandria, Va	6.2	6.6	5.1
All Saints, S. C	3.4	3.3	2.7
Amherst, Mass	5.0	5.0	4.3
Angelica, N.Y	7.3	5.0	4.3
Ann Arbor, Mich.	....	....	....
(Winchell.)	6.8	6.0	4.6
Ann Arbor, " (Woodruff)	7.0	6.3	5.8
Ashland, Va	4.8	5.1	3.5
Athens, Ill	5.6	5.9	4.1
Auburn, Ala	4.0	3.9	2.0
Augusta, Ill	5.8	4.7	3.7
Austin, Texas	6.3	5.6	2.8
Baldwin's Inst., Ohio	5.8	5.5	5.4
Baldwinsville, N. Y	7.8	6.5	5.8
Ballardsville, Ky	5.2	5.8	3.8
Battle Creek, Mich	5.8	5.5	4.3
Bedford, Pa	5.6	6.8	6.0
Beloit, Wis	3.9	2.9	2.9
Beverly, N. Y	5.4	5.8	5.1
Bladensburg, Md	5.5	3.0	3.3
Bloomfield, N. J	4.7	5.3	4.2
Boston, Mass.(1)	....	....	....
Brandon, Vt	6.0	5.9	4.8
Burlington, N. J	5.4	4.9	5.5
Burlington, Vt	2.9	3.1	2.8
Camden, S. C	5.9	5.3	5.2
Canton, N. Y	6.7	6.7	5.0
Carmel, Me	4.8	3.9	4.4
Castleton, Vt	6.6	6.5	6.0
Cedar Keys, Fa	4.3	4.0	3.2
Ceresco, Wis	3.0	3.3	3.7
Chapel Hill, N. C	4.4	4.1	5.3
Cincinnati, Ohio.(2)	....	....	....
Crichton's Store, Va	5.5	5.4	5.7
Concord, N. H	4.2	4.5	3.7
Cooper, Mich	6.1	5.5	5.0
Danville, Ky	4.2	4.1	3.6
Deaf & Dumb Inst., N. Y.	6.4	6.0	4.0
Detroit, Mich	9.0	7.7	7.8
Dubuque, Iowa	4.9	5.1	3.8
Easton, Pa	5.6	6.3	4.7
Exeter, N. H	5.2	4.6	3.6
Fort Madison, Iowa	5.9	4.2	4.1
Frederick, Md	6.5	6.1	5.3
Friendship, Tenn	4.0	2.7	3.4
Fryeburg, Me	4.9	4.2	3.6
Gardiner, Me	6.1	6.1	4.9
Garlandsville, Miss	4.4	4·2	4.3
Germantown, Ohio	6.7	6.7	4.8
Gettysburg, Pa	5.2	6.2	4.5
Glenwood, Tenn	4.9	5.1	4.2
Gouverneur, N. Y	7.5	7.9	6.8
Grand Rapids, Mich	6.1	4.8	4.2
Granville, Ohio	5.5	6.5	6.1
Green Springs, Ala.	4.2	3.7	2.5
Great Falls, N. H	6.3	5.2	3.8
Hannibal, Mo	5.9	4.7	3.4
Harrisburg, Pa	4.9	5.8	4.8
Hillsboro, Ohio	4.9	5.5	....
Jackson, Ohio	5.5	5.3	4.9
Jackson, O., (Crookham).	5.8	6.4	6.3
Jacksonville, Fa	3.6	3.7	1.9
Key West, Fa	4.7	3.7	....
Knox Hill, Fa	6.9	5.1	7.5
Knoxville, Tenn.	4.2	5.2	3.5
Lacquiparle, Minn	3.7	4.3	2.6
Lebanon, Tenn.	4.2	4·3	3.6
Lewisburg, Va	6.4	7.1	6.0
Lima, Pa	6.1	6.3	5.9
Lodi, N. Y	8.0	7.3	6.4
Londonderry, N. H	4.4	4.3	4.2
Lowville, N. Y	6.5	6.2	6.2
Madison, Ind	6.6	5.7	6.3
Manchester, Ill	5.3	5.8	4.0
Manchester, N. H	5.6	6.0	5.1
Marietta, Ohio	4.5	5.0	5.0
Mendon, Mass	6.0	5.6	4.2
Millersburg, Ky	5.8	5.2	3.9
Milton, Ind	5.2	5.3	4.3
Milwaukie, Wis	4.8	4.7	4.5
Monroe, Mich	4.9	5.2	6.2
Morrisville, Pa	4.4	4.5	4·2
Moss Grove, Pa	6.5	6.3	5.5
Mt. Vernon, Va	5.6	5.2	5.1
Muscatine, Iowa	4.6	4.5	2.6
Nantucket, Mass	5.8	6.0	4.0
Newark, Ohio	6.1	6.3	5.6
New Bedford, Mass.	6.1	7.1	4.6
Newburyport, Mass	5.2	5.8	3.4
New Harmony, Ind	5.4	5.9	4.4
New Lisbon, Ohio	5.4	5.3	5.2
New London, Ct	5.0	5.5	4.9
New Orleans, La	6.5	5.3	3.5
New Wied, Texas	5.5	6.0	7.5
Norristown, Pa	5.6	5.6	4.2
North Attleboro, Mass	5.5	5.4	3.9
Oberlin, Ohio	5.2	4.6	5·0
Oldtown, Me	2.4	2.5	2.7
Ottawa, Ill	5.2	4.8	4.5
Oxford, Miss	4.0	3.8	4.3
Penn Yan, N. Y	8.4	6.2	7.3
Perry, Me	6.2	4.0	5.5
Philadelphia, Pa	6.7	6.7	5.7
Platteville, Wis	5.1	5.5	4.5
Pocopson, Pa	6.1	6.6	5.5
Pomfret, Conn	5.2	5.2	4.1
Pottsville, Pa	5.8	6.4	5.5
Poultney, Iowa	6.0	6.1	5.1
Princeton, Mass	5.7	5.6	4.5
Providence, R. I	5.6	5.5	3.3
Quasqueton, Iowa	5.2	4.8	4.6
Randolph, Pa	7.6	7.2	6.8

* For changes in hours of observations see tables of Thermometer for this month.
(1) Mean for the month 3.5.
(2) Mean for the month 3.5.

NAME OF STATION.	7 A. M.	2 P. M.	9 P. M.	NAME OF STATION.	7 A. M.	2 P. M.	9 P. M.
Richmond, Mass	6.4	5.7	5.5	Springdale, Ky	5.1	5.9	5.8
Richmond, Ohio	5.6	5.6	5.7	Springfield, Mass	4.5	4.3	3.8
Sag Harbor, N. Y	5.6	5.2	4.8	Steuben, Me	6.5	5.8	6.7
St. James, Mich	6.4	5.5	5.3	Tuscaloosa, Ala	4.5	3.6	4.6
St. Johnsbury, Vt	3.8	2.4	2.9	Upper Alton, Ill	4.1	3.6	2.7
St. Louis, Mo.	5.3	4.1	4.4	Urbana, Ohio	6.2	6.2	5.2
St. Martin, Ca., E	4.0	4.5	4.9	Wampsville, N. Y	8.0	6.6	7.4
Saugatuck, Mich	7.2	6.7	5.3	Warrington, Fa	4.9	3.8	3.1
Savannah, Ga.	4.5	3.2	3.3	West Haverford, Pa	5.5	6.4	....
Saybrook, Ct	5.5	6.0	4.7	Whitemarsh Isl'd, Ga	4.0	3.4	3.5
Schellman Hall, Md	5.1	5.2	4.7	Williamstown, Mass	7.5	6.6	5.5
Smithville, Pa	6.8	6.5	6.6	Winchester, Va	4.9	4.4	5.0
Southwick, Mass	5.9	6.6	2.0	Woods Hole, Mass	5.5	5.7	5.2
Sparta, Ga	5.2	4.9	3.3	Worcester, Mass.	4.5	4.8	2.7
Spencertown, N. Y	5.3	4.9	3.8	Zanesville, Ohio	6.0	6.7	6.5
....	....	....	....	....	....	....	....

NAME OF STATION.	7 A. M.	2 P. M.	9 P. M.	NAME OF STATION.	7 A. M.	2 P. M.	9 P. M.
Alexandria, Va.	5.0	4.0	3.2	Hillsboro, Ohio	3.5	3.2	....
All Saints, S. C	3.0	3.3	2.5	Jackson, O., (Crookham).	6.1	6.5	3.8
Amherst, Mass	4.6	5.5	4.8	Jackson, O, (Wood)	4.3	3.7	3.5
Angelica, N. Y.	5.3	3.2	2.9	Jacksonville, Fa	3.6	3.0	1.2
Ann Arbor, Mich	....	....	....	Key West, Fa	3.6	6.5	....
(Woodruff)	4.3	3.7	4.0	Knox Hill, Fa	5.9	4.2	2.6
Ann Arbor, " (Winchell)	4.6	4.1	4.2	Knoxville, Tenn	5.0	5.2	2.6
Ashland, Va	2.3	2.4	1.5	Lacquiparle, Minn	3.2	4.2	2.3
Athens. Ill	3.9	4 8	4.5	Lewisburg, Va	4.4	5.4	3.8
Auburn, Ala	2.5	3.0	1.1	Lima, Pa	5.8	5.5	4.0
Augusta, Ill	4.6	2.5	4.0	Lodi, N. Y	6.7	5.9	4.6
Austin, Texas	7.3	5.2	2.9	Londonderry, N. H.	4.4	5.6	5.7
Baldwinsville, N. Y.	5.4	6.2	4.6	Lowville, N. H	4.8	5.3	5.2
Barnstable, Mass.	4.7	6.3	5.5	Madison, Wis	5.4	4.6	4.4
Battle Creek, Mich	3.0	3.0	4.2	Manchester, Ill	5.6	4.4	3.4
Bedford, Pa	4.6	4.4	3.2	Manchester, N. H	5.8	5.7	5.8
Beloit, Wis	1.9	2.0	2.0	Marietta, Ohio	3.3	3.4	2.6
Beverly, N. Y	4.4	6.0	3.6	Mendon, Mass	5.2	6.4	5.8
Bladensburg, Md	5.4	3.5	2.0	Millersburg, Ky.	4.4	3.9	2.7
Bloomfield, N. J	3.7	4.6	4.3	Milton, Ind	3.0	1.9	2.6
Brandon, Vt	5.7	5.5	4.5	Milwaukie, Wis	2.8	3.0	2.3
Burlington, N. J	4.3	3.3	3.8	Monroe, Mich	3.8	4.6	5.8
Burlington. Vt	3.3	3.3	2.6	Morrisville, Pa	3.4	3.7	2.9
Camden, S. C	5.8	5.8	5.1	Moss Grove, Pa	3.9	4.1	3.0
Canton, N. Y	5.7	6.3	5.0	Mt. Vernon, Ohio	2.6	3.0	3.2
Carmel, Me	4.3	5.2	6.2	Muscatine, Iowa	2.5	3.9	3.5
Cedar Keys, Fa	1.7	2.3	1.4	Nantucket, Mass	4.3	4.8	5.8
Ceresco, Wis	5.5	4.7	5.4	Newark, Ohio.	3.6	4.6	3.0
Chapel Hill, N. C	3.4	3.9	3.6	New Bedford, Mass	5.3	6.7	7.5
Concord, N. H	4.4	5.0	4.9	Newburyport, Mass	4.9	5.3	5.5
Cooper, Mich	3.7	2.6	3.7	New Harmony, Ind	5.8	4.3	3.9
Craftsbury, Vt.	6.1	5.6	5.1	New Lisbon, Ohio	3.3	3.1	2.2
Crichton's Store, Va	4.6	3.8	5.5	New London, Ct	4.6	4.2	5.0
Danville, Ky	3.4	3.2	2.4	New Orleans, La.	3.7	3.9	1.4
Deaf and Dumb Inst., N.Y.	4.9	5.6	4.1	New Wied, Texas	1.6	6.4	5.5
Detroit, Mich	7.7	7.9	7.4	Norristown, Pa	4.7	4.4	4.4
Dubuque, Iowa	4.0	4.1	3.0	North Attleboro, Mass	4.5	5.8	4.5
Easton, Pa	4.8	4.8	4.0	Oberlin, Ohio	3.2	3.1	3.3
Exeter, N. H	4.2	5.3	5.0	Oldtown, Me	2.4	2.4	5.2
Fort Madison, Iowa	3.9	4.0	4.1	Ottawa, Ill	4.7	4.1	4.8
Frederick, Md	6.2	5.0	4.3	Oxford, Miss	3.2	3.5	2.9
Friendship, Md	4.2	3.0	2.6	Penn Yan, N. Y	6.6	5.2	5.2
Fryeburg, Me.	5.0	4.9	5.0	Perry, Me	5.1	4.7	6.8
Gardiner, Me	5.2	6.7	5.6	Philadelphia, Pa	6.9	6.2	4.5
Garlandsville, Miss	4.4	4.4	4.0	Platteville, Wis	3.9	4.3	3.1
Germantown, Ohio	5.1	5.7	4.0	Pocopson, Pa	6.4	5.3	4.5
Gettysburg, Pa	4.9	3.8	2.5	Pomfret, Ct.	4.1	5.7	5.2
Glenwood, Tenn	6.4	5.8	4.9	Pottsville, Pa	6.5	5.5	4.6
Gouverneur, N. Y.	5.9	6.8	4.9	Poultney, Iowa.	4.2	4.6	4.1
Grand Rapids, Mich	4.2	3.5	3.8	Princeton, Mass	5.3	6.5	5.6
Granville, Ohio	3.4	3.4	3.8	Providence, R. I.	4.5	6.4	5.1
Green Springs, Ala	2.9	2.9	0.9	Quasqueton, Iowa	4.0	4.8	3.0
Great Falls, N. H	4.3	4.9	4.7	Randolph, Pa.	5.1	4.9	3.5
Hannibal, Mo	4.4	5.2	4.7	Richmond, Mass	6.1	6.0	5.3
Harrisburg, Pa	5.2	4.2	3.4	Richmond, Ohio.	3.8	2.8	2.9
....	....	....	....	....	....	....	....

NAME OF STATION.	7 A. M.	2 P. M.	9 P. M.	NAME OF STATION.	7 A. M.	2 P. M.	9 P. M.
Sag Harbor, N. Y.........	4.4	4.8	4.5	Spencertown, N. Y........	4.8	5.5	4.3
St. James, Mich.	4.4	5.0	4.6	Springfield, Mass..........	4.2	4.9	4.5
St. Johnsbury, Vt.........	3.9	3.1	3.3	Steuben, Me................	6.2	6.3	5.9
St. Louis, Mo...............	4.7	3.9	4.2	Thornbury, N. C...........	5.0	5.6	3.5
St. Martin, Ca., East. ...	4.5	4.3	5.2	Unionville, Ohio...........	4.6	4.1	4.9
Saugatuck, Mich.	4.1	4.4	3.7	Upper Alton, Ill...........	3.2	4.2	1.8
Savannah, Ga.............	3.9	3.2	2.5	Wampsville, N. Y	6.0	6.1	5.0
Savannah, Ohio............	4.3	4.2	4.3	Warrington, Fa............	3.1	1.9	2.1
Saybrook, Ct................	3.8	5.5	4.5	West Haverford, Pa.......	4.0	4.6	....
Schellman Hall, Md......	4.8	3.8	3.1	Whitemarsh Isl'd, Ga....	2.9	3.2	2.6
Smithsonian Inst. (1)......	6.0	4.4	3.1	Williamstown, Mass......	5.9	7.1	5.7
Smithville, N. Y..........	4.8	5.4	4.5	Winchester, Va............	5.1	4.2	4.8
Southwick, Mass.........	4.5	5.2	6.0	Worcester, Mass...........	4.5	4.2	4.1
Sparta, Ga.....	5.7	4.6	1.9	Zanesville, Ohio...........	5.4	5.4	5.0

(1) No obs. for the first 10 days of the month.

NAME OF STATION.	7 A. M.	2 P. M.	9 P. M.	NAME OF STATION.	7 A. M.	2 P. M.	9 P. M.
Alexandria, Va.	3.9	4.0	2.9	Jackson, O., (Crookham).	5.4	7.0	5.9
All Saints, S. C	4.2	4.3	3.6	Jackson, Ohio, (Gilman).	4.5	4.0	5.9
Amherst, Mass.	4.4	4.0	3.8	Jacksonville, Fa.	4.0	3.9	1.5
Angelica, N. Y.	2.3	2.1	3.1	Key West, Fa.	4.2	4.0	....
Ann Arbor, Mich.	....	....	....	Knox Hill, Fa.	5.9	6.7	2.8
(Winchell)	3.7	4.2	2.9	Knoxville, Tenn.	4.0	4.6	3.2
Ann Arbor, " (Woodruff)	4.0	3.7	3.1	Lacquiparle, Min.	3.1	3.5	4.0
Ashland, Va.	3.3	4.0	4.5	Lewisburg, Va.	6.8	6.0	5.7
Athens, Ill.	4.8	4.7	4.5	Lima, Pa.	5.4	5.1	4.3
Auburn, Ala.	3.5	4.5	1.6	Lodi, N. Y.	4.4	5.0	3.9
Augusta, Ill.	5.2	4.9	4.2	Lowville, N. Y.	3.6	3.9	3.7
Austin, Texas	8.4	6.2	4.3	Madison, Wis.	4.0	3.7	3.7
Baldwinsville, N. Y.	3.7	3.7	2.5	Madison, Ind.	5.0	4.8	4.6
Battle Creek, Mich.	3.0	2.6	2.7	Madrid, N.Y.	2.3	3.1	3.4
Bedford, Pa.	3.9	4.5	3.3	Manchester, Ill.	6.5	6.0	5.0
Beloit, Wis.	2.7	2.6	3.2	Manchester, N. H.	5.2	6.2	4.6
Beverly, N. Y.	4.5	4.9	4.1	Marietta, Ohio.	3.9	4.5	3.4
Bladensburg, Md.	4.2	4.0	2.9	Meadville, Pa.	3.4	3.6	3.8
Bloomfield, N. J.	3.6	3.8	3.3	Mendon, Mass.	3.2	4.0	4.2
Boston, Mass..(1)	....	....	....	Millersburg, Ky.	4.0	4.6	4.3
Brandon, Vt.	4.1	3.4	3.1	Milton, Ind.	3.3	3.6	3.5
Burlington, Vt.	2.2	1.9	2.3	Milwaukie, Wis.	4.1	3.9	3.9
Burlington, N. J.	3.3	2.6	2.5	Morrisville, Pa.	2.9	2.6	3.3
Camden, S. C.	5.2	6.0	5.5	Moss Grove, Pa.	4.0	3.5	3.8
Canton, N. Y.	4.0	4.2	3.6	Muscatine, Iowa.	3.7	2.9	2.7
Carmel, Me.	5.1	6.5	4.3	Nantucket, Mass.	4.7	3.9	5.1
Cedar Keys, Fa.	4.0	3.4	2.3	Newark, Ohio	4.7	5.0	4.6
Ceresco, Wis	5.5	4.6	5.7	New Bedford, Mass.	7.1	7.0	7.3
Chapel Hill, N. C.	5.5	4.1	6.3	Newburyport, Mass.	6.1	5.3	5.2
Cincinnati, Ohio.(2)	....	....	....	New Harmony, Ind.	5.5	5.4	3.7
Cleveland, Ohio.	4.4	4.7	3.7	New Lisbon, Ohio	2.5	3.9	3.2
Concord, N. H.	3.4	3.5	2.8	New London, Ct.	4.4	4.3	4.5
Cooper, Mich.	2.0	2.2	2.3	New Orleans, La.	3.5	4.1	1.2
Craftsbury, Vt.	4.8	4.8	3.1	New Wied, Texas.	0.8	7.0	7.2
Crichton's Store, Va.	4.6	4.2	5.4	Norristown, Pa.	4.8	5.0	4.9
Deaf and Dumb Inst., N.Y.	3.8	4.1	4.5	North Attleboro, Mass.	4.9	4.4	3.8
Detroit, Mich	5.9	4.8	2.8	Oberlin, Ohio	2.9	2.8	4.2
Dubuque, Iowa.	4.9	4.2	3.8	Ottawa, Ill	4.6	5.7	4.2
Easton, Pa.	3.5	4.8	3.5	Oxford, Miss.	4.0	4.9	3.3
Exeter, N. H.	4.5	4.4	4.7	Penn Yan, N. Y.	4.9	4.3	6.5
Fort Madison, Iowa.	4.5	5.0	3.9	Perry, Me.	6.3	6.8	7.8
Frederick, Md.	4.3	4.0	3.2	Philadelphia, Pa.	6.1	6.8	5.0
Friendship, Tenn.	4.7	3.3	2.7	Platteville, Wis.	4.0	3.7	4.3
Fryeburg, Me.	4.4	3.8	4.7	Pocopson, Pa.	5.4	6.2	3.9
Gardiner, Me.	6.5	6.8	5.1	Pomfret, Conn.	5.1	4.1	4.9
Garlandsville, Miss.	3.8	5.9	3.9	Pottsville, Pa.	4.3	5.3	3.7
Gettysburg, Pa.	3.9	4.5	3.5	Poultney, Iowa.	5.0	4.6	4.1
Glenwood, Tenn.	6.3	5.5	3.8	Princeton, Mass.	5.8	5.9	3.7
Gouverneur, N. Y.	3.4	4.4	2.5	Providence, R. I.	5.1	4.8	4.5
Grand Rapids, Mich.	3.4	2.8	2.8	Quasqueton, Iowa.	4.2	5.1	4.5
Granville, Ohio.	4.4	4.7	5.2	Randolph, Pa.	4.3	4.6	4.1
Green Springs, Ala.	3.5	4.6	1.6	Richmond, Mass.	4.3	5.2	5.1
Great Falls, N. H.	5.1	5.1	3.8	Richmond, Ohio.	3.0	3.6	4.5
Hannibal, Mo.	6.5	5.1	4.3	Sag Harbor, N. Y.	4.5	4.2	3.6
Harrisburg, Pa.	4.1	4.0	4.0	St. James, Mich.	3.5	3.9	4.8
Hillsboro, Ohio.	3.0	3.9	....	St. Johnsbury, Vt.	1.9	1.7	2.1

* For changes in hours of observation see notes for tables of Thermometer for this month.

(1) Mean for the month, 3.5.

(2) Mean for the month, 3.5.

NAME OF STATION.	7 A. M.	2 P. M.	9 P. M.	NAME OF STATION.	7 A. M.	2 P. M.	9 P. M.
St. Louis, Mo...............	6.1	5.4	4.6	Springfield, Mass.	3.0	3.0	3.3
St. Martin, Ca., E.........	2.2	2.3	1.4	Steuben, Me.	6.1	7.2	6.7
Saugatuck, Mich...........	4.0	3.5	3.4	Upper Alton, Ill...........	5.2	4.4	4.0
Savannah, Ga...............	3.7	4.3	3.0	Unionville, Ohio...........	4.4	3.8	4.6
Savannah, Ohio............	4.1	4.5	3.2	Urbana, Ohio................	4.8	5.6	4.2
Saybrook, Ct................	3.8	4.3	4.3	Wampsville, N. Y.........	3.6	4.5	3.1
Schellman Hall, Md........	3.6	3.6	3.0	Warrington, Fa.............	3.0	3.0	2.8
Smithsonian Inst.	5.0	4.1	3.5	West Haverford, Pa........	3.6	4.8	2.0
Smithville, N. Y...........	3.5	2.7	3.0	Whitemarsh Isl'd, Ga.....	3.4	4.5	2.4
Southwick, Mass(1)........	4.8	3.7	6.0	Williamstown, Mass........	3.8	4.3	3.9
Sparta, Ga....................	3.5	4.1	3.4	Winchester, Va.............	4.1	4.2	5.2
Spencertown, N. Y........	2.8	4.1	3.9	Worcester, Mass............	3.3	4.0	3.4
Springdale, Ky..............	4.4	5.1	4.3	....	....	....	....

(1) Obs. irregular, only three at 9 P. M.

NAME OF STATION.	7 A. M.	2 P. M.	9 P. M.	NAME OF STATION.	7 A. M.	2 P. M.	9 P. M.
Alexandria, Va	5.5	6.0	5.1	Jackson, O., (Crookham)	6.0	8.0	5.6
All Saints, S. C	3.5	4.2	4.1	Jackson, O., (Gilman)	5.0	6.1	5.3
Amherst, Mass	3.9	4.7	3.0	Jacksonville, Fa	3.9	5.4	3.4
Ann Arbor, Mich	....	....	....	Key West, Fa	5.9	4.8	....
(Woodruff)	5.2	7.0	4.6	Knox Hill, Fa	6.5	6.7	4.9
Ann Arbor, " (Winchell)	5.1	7.1	4.1	Knoxville, Tenn	4.5	5.5	3.9
Angelica, N. Y	7.0	5.9	6.0	Lewisburg, Va	6.3	6.9	4.3
Ashland, Va	3.9	4.7	3.0	Lima, Pa	5.7	6.3	5.4
Athens, Ill	4.3	3.3	2.6	Lodi, N. Y	7.7	7.1	6.5
Auburn, Ala	4.6	4.2	2.9	Londonderry, N. H	6.0	5.5	4.9
Augusta, Ill	4.0	3.7	2.5	Lowville, N. Y	6.0	5.7	4.2
Austin, Texas	6.0	5.7	2.6	Madison, Ind	6.3	6.4	4.3
Baldwinsville, N. Y	7.1	6.5	6.3	Madrid, N. Y	7.3	3.7	5.6
Battle Creek, Mich	4.4	5.1	3.4	Manchester, Ill	5.6	5.4	2·8
Bedford, Pa	5.1	7.2	4.2	Manchester, N. H	6 7	7.7	4.9
Beloit, Wis	3.9	3.4	2.8	Meadville, Pa	6.8	6.4	5.9
Beverly, N. Y	6.4	5.7	4.5	Mendon, Mass	5.4	5.9	5.1
Bladensburg, Md	5.0	6.1	7.2	Millersburg, Ky	4.7	5.0	3.9
Bloomfield, N. J	5.2	5.2	4.3	Milton, Ind	4.2	5.4	3.3
Boston, Mass. (1)	....	....	....	Milwaukie, Wis. (Winkler)	5.2	5.4	3.3
Brandon, Vt	6.4	5.4	4.5	Monroe, Mich. (Whelpley)	6.6	3.4	7.7
Burlington, Vt	4 0	3.5	3.4	Morrisville, Pa	4.7	3.3	4.8
Burlington, N. J	8.5	6.2	7.0	Moss Grove, Pa	6.9	7.0	5.2
Camden, S. C	5.3	5.1	4.6	Muscatine, Iowa	2.5	3.7	3.2
Canton, N. Y	7.4	6.6	5.3	Nantucket, Mass	2.7	2.8	2.9
Carmel, Me	6.1	6.6	4.6	New Bedford, Mass	6.4	6.5	5.6
Cedar Keys, Fa	3.4	4.7	2.1	Newburyport, Mass	7.6	6.2	7.6
Chapel Hill, N. C	5.6	4.1	6.0	New Harmony, Ind	3.6	5.9	3.4
Cincinnati, Ohio. (2)	....	....	....	New London, Ct	5.3	5.4	5.2
Cleveland, Ohio	6.6	7.0	5.6	New Orleans, La	4.8	6.4	1.2
Concord, N. H	5.4	5.2	3.5	New Wied, Texas	4.2	5.3	7.6
Cooper, Mich	5.1	4.2	4.5	Norristown, Pa	5.0	6.0	5.3
Craftsbury, Vt	6.7	6.9	4.4	North Attleboro, Mass	5.0	5.1	4.5
Crichton's Store, Va	4.4	5.1	4.4	Oberlin, Ohio	5.5	5.7	4.4
Danville, Ky	3.0	3.5	3.1	Oldtown, Me	2.4	3.0	2.1
Deaf & Dumb Inst N. Y	5.8	4.9	5.3	Ottawa, Ill	5.2	6.9	3.7
Delaware College	5.0	5.4	3.5	Oxford, Miss	4.2	5.1	2.3
Detroit, Mich	7.6	8.1	7.8	Penn Yan, N. Y	8.4	7.2	7.0
Dubuque, Iowa	4.7	5.2	3.5	Perry, Me	5.7	4.7	5.6
Easton, Pa	5.3	6.7	5.5	Philadelphia, Pa	6.3	7.2	6.4
Exeter, N. H	4.9	5.3	4.2	Platteville, Wis	3.5	5.5	4.0
Ft. Madison, Iowa	3.8	3.3	3.0	Pocopson, Pa	6.3	7.1	6.4
Frederick, City, Md	5.4	5.6	5.7	Pomfret, Conn	5.9	5.5	5.1
Friendship, Tenn	3.6	3.9	2.8	Pottsville, Pa	5.9	6.9	6.6
Fryeburg, Me	5.2	6.0	4.7	Poultney, Iowa	5.4	6.3	6.0
Gardiner, Me	7.4	7.2	5.4	Princeton, Mass	6.1	6.6	5.1
Germantown, Ohio	6.4	7.1	5 3	Providence, R. I	5.5	5.5	4.6
Gettysburg, Pa	4.7	6.6	5.3	Quasqueton, Iowa	3.9	5.1	4.4
Glenwood, Tenn	4.6	5.9	3.8	Randolph, Pa	6.4	7.3	5.0
Gouverneur, N. Y	6.5	7.5	7.4	Red River Settlement	3.4	3.5	3.5
Grand Rapids, Mich	5.2	6.0	3.1	Richmond, Ohio	5.6	6.5	4.8
Granville, Ohio	5.8	6.7	3.3	Sag Harbor, N. Y	5.3	5.4	4.6
Great Falls, N. H	5.5	5.9	4.8	St. James, Mich	4.8	4.7	4.9
Green Springs, Ala	2.5	4.0	0.8	St. Johnsbury, Vt	4.6	4.1	3.4
Harrisburg, Pa	4.8	5.5	4.8	St. Louis, Mo	5.0	5.0	2.6
Hillsboro, Ohio	5.3	4.9	....	St. Martin, Ca., E	5.1	5.3	4.9

* For changes in hours of observation see tables of Thermometer for this month.
(1) Mean for the month 6.5.
(2) Mean for the month 5.0.

NAME OF STATION.	7 A. M.	2 P. M.	9 P. M.	NAME OF STATION.	7 A. M.	2 P. M.	9 P. M.
Saugatuck, Mich..........	6.0	5.7	3.5	Steuben, Me.	4.6	6.9	5.8
Savannah, Ohio	6.7	7.2	5.1	Superior, Wis..............	3.0	4.6	4.6
Savannah, Ga.	3.6	5.3	3.5	Unionville, Ohio...........	6.7	6.6	5.7
Saybrook, Ct..............	5.3	5.5	4.2	Upper Alton, Ill.	3.2	3.6	1.9
Schellman Hall, Md........	4.8	5.5	5.2	Urbana, Ohio..............	6.3	6.6	4.8
Smithsonian Inst..........	5.4	4.8	5.6	Wampsville, N. Y.........	7.2	6.5	7.2
Smithville, N. Y...........	6.8	5.5	5.0	Warrington, Fa............	3.2	3.5	3.5
Southwick, Mass...........	6.4	6.7	5.0	West Haverford, Pa.......	4.6	6.6	....
South Thomaston, Me....	3.5	3.7	3.9	Whitemarsh Island, Ga. ..	3.6	4.1	3.3
Sparta, Ga.................	3.7	6.4	3.3	Williamstown, Mass.......	6.8	6.9	5.1
Spencertown, N. Y........	6.0	5.6	3.5	Winchester, Va............	4.7	5.5	5.9
Springdale, Ky............	3.8	5.2	5.0	Worcester, Mass..........	5.0	4.5	4.8
Springfield, Mass.	5.8	5.5	3.9	Zanesville, Ohio..........	5.4	6.7	5.2
....	....	....	....	....	....	....	....

NAME OF STATION.	7 A. M.	2 P. M.	9 P. M.	NAME OF STATION.	7 A. M.	2 P. M.	9 P. M.
Alexandria, Va	5.0	5.0	3.7	Jackson, Ohio, (Gilman).	4.3	5.2	5.1
All Saints, S. C	2.9	4.4	2.9	Jacksonville, Fa	3.8	4.8	2.6
Amherst, Mass	6.3	5.7	5.4	Key West, Fa	5.7	6.0	....
Ann Arbor, Mich	....	....	....	Knox Hill, Fa	5.8	7.3	6.4
(Woodruff)	5.7	6.0	4.2	Lcaquiparle, Min	3.1	4.2	2.4
Ann Arbor, " .(Winchell)	5.2	6.1	4.1	Lewisburg, Va	6.5	7.9	5.9
Arcola, Ohio.	6.1	5.4	4.9	Lima, Pa	5.5	5.6	5.5
Ashland, Va.	4.8	4.5	3.1	Lodi, N. Y	6.0	7.0	5.2
Athens, Ill	4.5	4.7	4.5	Lowville, N. Y	5.9	5.3	5.6
Augusta, Ill	4.5	3.6	3.7	Manchester, Ill.	4.0	5.0	4.8
Austin, Texas	6.4	5.8	2.1	Manchester, N. H	5.3	6.7	5.1
Baldwinsville, N. Y	5.3	6.1	5.0	Mendon, Mass	6.5	6.1	6.2
Battle Creek, Mich	4.1	4.8	2.7	Milton, Ind	4.0	4.2	3.1
Bedford, Pa	5.6	6.3	5.5	Milwaukie, Wis. (Winkler)	5.6	4.1	4.2
Beloit, Wis	3.8	2.9	2.1	Monroe, Mich. (Perkins).	5.0	6.3	7.4
Beverly, N. Y	7.0	5.1	5.7	Morrisville, Pa	4.4	2.7	4.8
Bladensburg, Md..(1)	4.0	4.5	3.5	Moss Grove, Pa	6.2	6.5	4.4
Bloomfield, N. J	6.2	5.8	5.6	Muscatine, Iowa.	3.7	3.5	2.7
Boston, Mass. (2)	....	....	....	Nantucket, Mass	6.1	5.2	5.6
Brandon, Vt	5.1	5.3	5.3	New Bedford, Mass	4.0	6.9	6.9
Burlington, Vt	2.8	2.4	2.6	Newburyport, Mass	8.4	5.0	7.0
Burlington, N. J	7.7	6.2	7.3	New Harmony, Ind	4.9	5.3	4.7
Camden, S. C	5.4	6.1	5.4	New Lisbon, Ohio	6.4	5.3	6.8
Canton, N. Y. (1)	6.3	5.3	3.6	New Orleans, La	3.5	7.4	2.1
Carmel, Me	5.8	5.5	5.2	New Wied, Texas	3.7	6.5	7.5
Cedar Keys, Fa	4.5	5.4	4.5	Norristown, Pa	5.9	6.7	5.5
Chapel Hill, N. C	5.2	4.2	6.4	North Attleboro, Mass	6.0	5.8	4.5
Chestertown, Md	5.7	5.0	6.1	Oberlin, Ohio	4.1	4.3	3.2
Cincinnati, Ohio (3)	....	....	....	Oldtown, Me.. (1)	1.3	1.2	1.3
Cleveland, Ohio	5.7	5.6	5.9	Ottawa, Ill	3.9	5.7	3.9
Concord, N. H	4.9	4.1	3.8	Oxford, Miss	2.1	4.1	3.1
Cooper, Mich	3.9	5.1	3.1	Penn Yan, N. Y	6.9	6.3	6.0
Craftsbury, Vt	5.7	6.2	4.6	Perry, Me	6.3	4.0	4.8
Crichton's Store, Va	4.2	4.4	4.5	Philadelphia, Pa.	6.3	6.0	5.6
Danville, Ky	2.9	3.9	2.2	Platteville, Wis	4.7	4.4	4.2
Deaf and Dumb Inst., N.Y.	7.2	6.1	6.3	Pocopson, Pa	6.3	6.2	4.8
Detroit, Mich	5.6	6.9	2.6	Pomfret, Conn	6.5	5.9	6.2
Dubuque, Iowa	5.5	4.4	3.5	Pottsville, Pa.	5.9	6.4	5.4
Easton, Pa	4.6	6.2	4.3	Poultney, Iowa	5.7	5.6	4.1
Exeter, N. H	5.4	5.1	4.5	Princeton, Mass	6.4	7.0	5.9
Fort Madison, Iowa	4.1	3.5	4.4	Providence, R. I	6.0	5.9	5.0
Frederick City, Md	5.1	5.0	4.0	Quasqueton, Iowa	5.5	5.5	3.2
Friendship, Tenn	3.4	3.5	3.7	Red River Settlement	4.7	3.7	4.2
Fryeburg, Me	6.3	4.1	....	Richmond, Mass	6.3	5.6	5.8
Gardiner, Me	7.4	6.3	6.3	Richmond, Ohio	5.2	5.9	4.1
Germantown, Ohio.	5.2	6.4	4.4	Sag Harbor, N. Y	6.8	5.8	5.6
Gettysburg, Pa	5.7	5.2	4.7	St. Louis, Mo	5.1	4.1	3.6
Glenwood, Tenn. (1)	4.4	6.0	0.8	St. James, Mich	5.5	4.2	4.1
Gouverneur, N. Y. (1)	4.6	7.5	8.4	St. Johnsbury, Vt	3.1	2.5	3.1
Grand Rapids, Mich	4.9	4.8	3.6	St. Martin, Ca. E	3.4	4.9	3.7
Granville, Ohio	5.2	6.4	5.4	Saugatuck, Mich	5.5	4.8	3.5
Green Springs, Ala.. (1)	3.6	4.9	1.6	Savannah, Ohio	6.1	6.0	4.0
Harrisburg, Pa	5.0	4.5	4.0	Savannah, Ga	3.2	5.1	3.0
Hillsboro, Ohio	3.4	2.9	....	Saybrook, Conn	7.0	4.9	5.1
Jackson, O., (Crookham).	5.9	7.3	5.1	Schellman Hall, Md	4.6	5.2	3.5
....	....	....	....	....	....	....	....

* For changes in hours of observation see tables of Thermometer for this month.

(1) Incomplete.

(2) Monthly mean 3.0.

(3) Monthly mean 3.7.

NAME OF STATION.	7 A. M.	2 P. M.	9 P. M.	NAME OF STATION.	7 A. M.	2 P. M.	9 P. M.
Smithsonian Inst........	5.8	3.8	3.6	Upper Alton, Ill..........	3.3	2.9	2.9
Smithville, N. Y...........	3.5	6.0	4.6	Urbana, Ohio..............	5.5	6.9	4.2
Southwick, Mass...........	5.8	5.5	7.5	Wampsville, N. Y.........	7.0	6.8	4.8
Sparta, Ga..................	4.7	6.3	6.9	Warrington, Fa............	4.4	4.9	4.3
Spencertown, N. Y........	4.7	4.6	4.7	West Haverford, Pa......	4.7	5.5	....
Springdale, Ky.	4.0	5.3	3.9	Whitemarsh Island, Ga. ..	2.8	4.0	3.0
Springfield, Mass.........	5.8	4.6	4.8	Williamstown, Mass.......	6.0	6.3	7.9
Steuben, Me	7.0	4.9	7.3	Worcester, Mass...........	4.5	4.9	3.5
Superior, Wis..............	3.3	2.3	3.4	Zanesville, Ohio...........	5.2	5.9	5.9
....	......	......	......	....	......	......	......
....	......	......	......	....	......	......	......
....	......	......	......	....	......	......	......

NAME OF STATION.	7 A. M.	2 P. M.	9 P. M.	NAME OF STATION.	7 A. M.	2 P. M.	9 P. M.
Alexandria, Va	5.7	6.5	4.5	Jacksonville, Fa	3.1	4.1	2.7
All Saints, S. C	4.3	5.2	4.7	Key West, Fa	5.7	5.2	....
Amherst, Mass	4.5	4.6	3.4	Knox Hill, Fa..(1)	6.0	7.4	6.3
Angelica, N.Y	3.2	2.5	3.2	Lacquiparle, Minn	4.8	4.9	4.1
Ann Arbor, Mich.	....	....	....	Lewisburg, Va	4.3	4.2	4.9
(Woodruff)	4.5	5.5	3.1	Lima, Pa	5.4	6.6	4.4
Ann Arbor, " (Winchell.)	3.8	4.8	2.5	Lodi, N. Y	5.0	5.9	4.0
Ashland, Va	6.0	4.6	1.6	Londonderry, N. H	3.2	3.9	2.1
Athens, Ill	4.7	4.8	3.4	Madrid, N. Y	4.8	5.3	3.6
Auburn, Ala	5.6	6.4	4.1	Manchester, Ill	4.2	5.4	3.2
Augusta, Ill	4.4	4.3	2.4	Manchester, N. H..(1)	4.5	5.2	3.4
Austin, Texas	6.8	7.1	3.0	Meadville, Pa(1)	5.1	4.1	3.9
Baldwinsville, N. Y	3.5	4.6	3.9	Mendon, Mass	4.5	4.3	3.4
Battle Creek, Mich	3.6	4.0	1.9	Milton, Ind	3.4	3.7	2.2
Bedford, Pa	5.5	5.0	3.8	Milwaukie, Wis. (Winkler)	4.6	3.9	1.7
Beloit, Wis	3.4	2.1	1.8	Monroe, Mich. (Whelpley)	4.8	5.7	6.5
Beverly, N. Y	4.7	4.6	3.9	Morrisville, Pa	4.0	3.8	3.2
Bladensburg, Md..(1)	5.6	4.5	5.0	Moss Grove, Pa	5.8	5.4	3.2
Bloomfield, N. J	4.4	4.3	2.4	Muscatine, Iowa	5.4	4.2	3.6
Brandon, Vt	3.9	3.1	3.2	Nantucket, Mass	4.1	2.6	2.6
Burlington, Vt	2.5	2.0	2.5	New Bedford, Mass.	6.1	5.4	5.7
Burlington, N. J	7.6	6.0	6.8	Newburyport, Mass	7.3	4.2	6.5
Camden, S. C	6.0	6.3	6.3	New Harmony, Ind	4.8	5.5	4.4
Canton, N. Y	4.3	4.6	3.9	New Lisbon, Ohio	3.8	3.8	5.5
Carmel, Me (1)	2.3	3.0	1.9	New Orleans, La	4.3	7.1	3.3
Cedar Keys, Fa	4.4	4.6	3.3	New Wied, Texas	3.4	5.0	7.9
Chapel Hill, N. C	5.8	3.6	7.5	Norristown, Pa	5.4	5.6	3.6
Chestertown, Md	4.5	4.9	4.6	North Attleboro, Mass	3.4	3.8	3.3
Cleveland, Ohio	5.3	4.7	3.7	Oberlin, Ohio	3.1	3.1	2.6
Concord, N. H	2.5	1.9	3.1	Oldtown, Me	1.6	1.4	1.8
Cooper, Mich	3.6	3.8	2.3	Ottawa, Ill	4.0	5.1	2.5
Craftsbury, Vt	4.6	4.2	3.0	Oxford, Miss.(1)	4.2	5.5	2.0
Crichton's Store, Va	5.8	5.3	5.6	Penn Yan, N. Y	4.8	5.8	6.5
Dubuque, Iowa	6.5	3.6	2.3	Perry, Me	4.2	3.6	4.4
Deaf and Dumb Inst. N.Y.	4.9	5.5	3.6	Philadelphia, Pa	6.2	6.8	4.4
Detroit, Mich	1.0	4.0	1.2	Platteville, Wis	6.0	3.8	3.3
Easton, Pa	6.0	4.8	3.9	Pocopson, Pa	6.1	5.8	5.1
Exeter, N. H	3.1	3.5	3.3	Pomfret, Conn	4.6	4.5	4.1
Fort Madison, Iowa	4.5	4.2	2.4	Port Gibson, Mis..(1)	2.4	5.3	5.7
Frederick City, Md	6.5	6.4	4.9	Pottsville, Pa	5.9	5.6	4.3
Fryeburg, Me	3.1	3.1	....	Poultney, Iowa. .(1)	6.0	4.1	2.1
Gardiner, Me	5.4	4.6	2.7	Princeton, Mass	4.3	5.5	4.6
Germantown, Ohio	5.9	5.5	4.0	Red River Settlement	5.1	5.7	3.3
Gettysburg, Pa	6.2	5.6	4.2	Richmond, Mass	5.5	4.7	4.6
Glenwood, Tenn(1)	7.7	8.5	3.0	Richmond, Ohio	3.5	3.4	1.4
Gouverneur, N. Y	4.4	4.5	2.5	Sag Harbor, N. Y	3.9	4.5	3.9
Grand Rapids, Mich	3.9	5.0	2.1	St. James, Mich.	6.4	4.1	4.2
Granville, Ohio	4.6	5.3	3.5	St. Louis, Mo	5.2	5.9	3.3
Green Springs, Ala.	4.7	6.5	3.8	St. Martin, Ca., East.	2.3	3.3	3.7
Harrisburg, Pa	5.6	5.3	4.8	Saugatuck, Mich.	4.8	4.3	2.3
Hillsboro, Ohio.(2)	3.9	3.6	....	Savannah, Ohio	4.9	4.9	2.9
Jackson, O., (Crookham).	6.4	6.4	3.4	Savannah, Ga	4.3	6.4	4.7
Jackson, Ohio, (Gilmor)	4.6	5.9	4.2	Saybrook, Ct......(1)	3.1	4.5	3.2

* For changes in hours of observation see tables of Thermometer for this month.

(1) Record incomplete.

(2) Only one observation at 9 P. M.

NAME OF STATION.	7 A. M.	2 P. M.	9 P. M.	NAME OF STATION.	7 A. M.	2 P. M.	9 P. M.
Schellman Hall, Md......	5.2	5.0	3.4	Upper Alton, Ill	3.7	3.8	1.2
Smithsonian Inst.	7.1	5.7	3.8	Urbana, Ohio......	5.8	5.7	3.9
Smithville, N. Y...........	4.4	3.6	3.5	Wampsville, N. Y	4.2	4.5	4.8
Southwick, Mass..........	3.3	4.9	....	Warrington, Fa	3.3	3.8	4.0
Sparta, Ga.....	4.7	6.3	3.9	West Haverford, Pa.	4.6	6.0	....
Spencertown, N. Y........	3.7	3.5	3.5	Whitemarsh Isl'd, Ga. ...	4.0	5.3	3.7
Springdale, Ky.............	5.3	5.6	4.0	Williamstown, Mass	4.2	5.2	3.9
Springfield, Mass..........	3.5	4.1	2.5	Worcester, Mass......... ...	3.1	2.6	2.4
Steuben, Me................	4.6	5.2	4.9	Zanesville, Ohio..	5.2	5.2	4.5
Superior, Wis.............	4.3	4.3	4.5				

NAME OF STATION.	7 A. M.	2 P. M.	9 P. M.	NAME OF STATION.	7 A. M.	2 P. M.	9 P. M.
Alexandria, Va............	6.0	5.8	3.7	Jacksonville, Fa..........	5.3	4.7	4.1
All Saints, S. C...........	3.6	4.1	3.6	Key West, Fa...............	5.4	4.7	....
Amherst, Mass.............	4.9	4.3	3.0	Knox Hill, Fa.............	5.2	6.9	4.1
Angelica, N. Y............	4.5	2.4	2.9	Lacquiparle, Minn	5.3	5.2	4.7
Ann Arbor, Mich..........	....	....	....	Lewisburg, Va............	5.0	5.0	3.7
(Woodruff)...............	6.6	6.6	5.5	Lima, Pa....................	4.7	5.8	3.3
Ashland, Va...............	5.7	5.4	3.3	Lodi, N. Y..................	5.3	4.6	3.2
Athens, Ill..................	6.4	5.6	4.7	Londonderry, N. H.	3.4	2.7	2.9
Auburn, Ala................	4.0	5.7	1.7	Manchester, Ill (1).........	6.5	6.4	3.7
Augusta, Ill................	5.8	5.6	3.9	Manchester, N. H.........	4.9	4.8	3.9
Austin, Texas..............	7.2	6.9	2.6	Meadville, Pa..............	6.5	5.9	4.0
Baldwinsville, N. Y......	3.8	4.6	3.1	Mendon, Mass..............	4.5	4.4	2.5
Battle Creek, Mich........	5.7	5.9	5.0	Milton, Ind..................	5.0	5.5	4.8
Beloit, Wis..................	3.3	2.6	3.7	Milwaukie, Wis (Winkler)	5.3	6.1	4.7
Beverly, N. Y..............	5.3	5.2	3.5	Monroe, Mich (Whelpley)	6.5	6.8	7.0
Bladensburg, Md..........	4.9	5.4	4.7	Montreal, Ca...............	5.1	5.7	4.3
Bloomfield, N. J...........	3.6	3.9	2.2	Morrisville, Pa............	3.3	3.4	2.8
Boston, Mass. (1)..........	....	....	....	Moss Grove, Pa............	6.5	6.9	3.7
Bowling Green, Ky........	5.4	5.5	3.7	Muscatine, Iowa...........	6.9	5.2	5.1
Brandon, Vt.................	5.6	4.	3.5	Nantucket, Mass...........	3.2	3.1	2.8
Burlington, Vt.............	3.3	1.9	2.6	New Bedford, Mass.......	5.8	5.5	4.2
Burlington, N. J..........	8.6	6.7	8.4	Newburyport, Mass.......	6.7	5.3	5.8
Camden, S. C..............	5.4	4.9	3.7	New Harmony, Ind.......	5.5	6.4	5.1
Canton, N. Y..............	5.1	5.6	4.2	New London, Ct..........	3.9	3.2	1.9
Carmel, Me.................	5.0	5.2	3.0	New Orleans, La.	3.9	7.0	2.7
Cedar Keys, Fa...........	2.6	4.6	2.2	New Wied, Texas	3.6	5.2	9.5
Chapel Hill, N. C.........	2.9	3.8	1.4	Norristown, Pa............	3.8	4.9	2.8
Cincinnati, Ohio. (2).......	....	....	....	North Attleboro, Mass...	4.2	3.6	2.5
Cleveland, Ohio...........	6.6	6.9	4.0	Oberlin, Ohio	5.0	5.2	3.0
Concord, N. H.............	3.8	2.2	2.4	Ottawa, Ill	3.7	7.2	4.3
Cooper, Mich...............	4.4	3.9	4.8	Oxford, Miss...............	3.7	4.6	4.0
Craftsbury, Vt.	6.0	4.5	3.9	Penn Yan, N. Y...........	5.4	4.7	4.8
Crichton's Store, Va......	.6.3	3.9	3.4	Perry, Me..................	4.4	4.7	4.2
Danville, Ky...............	3.8	4.0	2.8	Philadelphia, Pa..........	5.3	5.7	2.7
Deaf and Dumb Inst., N.Y.	4.7	5.7	2.9	Platteville, Wis...........	7.0	6.4	5.5
Delaware College..........	6.9	5.5	2.9	Pocopson, Pa..............	5.9	7.0	3.5
Detroit, Mich..............	4.7	4.1	2.1	Pomfret, Ct................	4.3	4.1	3.1
Dubuque, Iowa...........	7.5	5.9	3.8	Pottsville, Pa..............	5.9	5.5	3.9
Easton, Pa..................	5.6	5.1	2.8	Poultney, Iowa	6.1	5.7	5.8
Exeter, N. H...............	4.2	3.7	2.3	Princeton, Mass...........	5.4	5.1	3.9
Fort Madison, Iowa.......	7.5	6.2	6.4	Providence, R. I..........	4.5	4.3	2.9
Frederick, Md..............	6.0	6.0	5.5	Red River Settlement....	4.2	4.6	4.7
Fryeburg, Me.	3.8	3.3	....	Richmond, Mass..........	4.1	3.5	2.5
Germantown, Ohio........	6.8	7.0	5.5	Richmond, Ohio...........	5.7	5.2	4.2
Gettysburg, Pa............	5.7	5 9	4.6	Sag Harbor, N. Y.........	4.4	4.4	2.4
Glenwood, Tenn...........	6.3	7.3	2.8	St. James, Mich...........	6.6	5.2	4.7
Gouverneur, N. Y.........	5.3	5.6	3.3	St. Louis, Mo...............	5.3	4.4	4.1
Grand Rapids, Mich......	6.3	5.8	3.7	St. Martin, Ca., E.........	3.7	4.2	4.3
Granville, Ohio............	6.1	5.4	5.2	Saugatuck, Mich..........	5.5	6.3	4.8
Harrisburg, Pa............	5.4	4.9	4.7	Savannah, Ohio...........	6.0	6.4	4.7
Hillsboro, Ohio...........	5.7	4.8	....	Savannah, Ga.	4.0	6.0	1.7
Jackson, Ohio. (Gilmor).	3.5	5.6	1.1	Saybrook, Ct...............	4.0	4.3	3.0
....	....	....	....	....	....	....	....

* For changes in hours of observation see tables of Thermometer for this month.

(1) Mean for the month 1.8.

(2) Mean for the month 4.6.

NAME OF STATION.	7 A. M.	2 P. M.	9 P. M.	NAME OF STATION.	7 A. M.	2 P. M.	9 P. M.
Schellman Hall, Md......	4.6	5.3	4.0	Upper Alton, Ill........	5.2	4.2	2.6
Smithsonian Inst..........	8.1	5.5	4.1	Urbana, Ohio..............	6.1	7.1	5.2
Smithville, N. Y..........	4.5	4.4	3.2	Wampsville, N. Y.........	5.2	5.3	3.1
Southwick, Mass..........	4.2	3.2	....	Warrington, Fa............	2.9	3.6	3.3
Sparta, Ga..................	4.2	5.4	2.8	West Haverford, Pa.......	4.3	5.4	....
Spencertown, N. Y........	3.2	3.4	3.6	Whitemarsh Isl'd, Ga.....	3.9	4.8	2.4
Springdale, Ky	5.1	5.5	5.3	Williamstown, Mass.......	6.0	5.4	4.1
Springfield, Mass..........	3.7	4.1	2.7	Worcester, Mass...........	4.3	4.2	3.1
Steuben, Me................	4.8	4.6	4.3	Zanesville, Ohio........ ...	5.7	5.9	5.5
Unionville, Ohio..........	6.3	5·0	4.9	....	....	....	....
....	....	....	....	....	....	....	....
....	....	....	....	....	....	....	....

NAME OF STATION.	7 A. M.	2 P. M.	9 P. M.
Alexandria, Va	4.4	4.5	3.9
All Saints, S. C	2.9	3.3	2.3
Amherst, Mass	6.4	6.0	5.4
Angelica, N. Y.	7.2	6.5	7.2
Ann Arbor, Mich.	....	....	....
(Woodruff.)	4.2	5.4	3.9
Ann Arbor, " (Winchell)	4.9	5.6	4.2
Ashland, Va.	4.3	3.6	2.2
Athens, Ill	4.2	3.9	3.3
Auburn, Ala.	3.3	3.5	1.8
Augusta, Ill	3.3	3.5	1.8
Austin, Texas	6.4	5.6	3.4
Baldwinsville, N. Y	6.7	7.1	5.5
Battle Creek, Mich	4.5	4.9	4.4
Bedford, Pa.	6.2	6.7	4.1
Beloit, Wis	2.6	2.3	1.8
Beverly, N. Y	5.2	6.6	5.7
Bloomfield, N. J.	5.2	5.3	4.5
Boston, Mass. (1)	....	....	....
Bowling Green, Ky	3.0	3.9	2.7
Brandon, Vt.	7.9	6.6	7.0
Burlington, Vt.	4.5	4.0	4.0
Burlington, N. J	7.2	6.5	6.6
Camden, S. C	4.3	4.5	2.9
Camden, Ark.	7.4	5.6	4.7
Carmel, Me.	8.2	6.8	6.0
Canton, N. Y	7.4	4.3	6.5
Castleton, Vt.	6.5	5.8	5.5
Cedar Keys, Fa	3.9	4.1	2.7
Chapel Hill, N. C	6.1	5.6	7.4
Chestertown, Md	4.5	3.9	3.7
Cincinnati, Ohio(2)	....	....	....
Cleveland, Ohio	6.5	6.4	6.5
Concord, N. H	5.2	6.0	4.6
Cooper, Mich	5.0	4.3	4.2
Craftsbury, Vt	7.1	6.9	7.1
Crichton's Store, Va	4.2	4.0	4.4
Deaf and Dumb Inst., N.Y.	4.4	5.8	4.7
Delaware College.	3.8	5.0	4.6
Detroit, Mich	4.6	4.4	3.9
Dubuque, Iowa	4.1	4.2	3.6
Easton, Pa	4.7	5.4	3.2
Exeter, N. H	5.9	6.2	5.5
Fort Madison, Iowa	3.5	3.1	3.5
Frederick City, Md	5.2	5.1	4.0
Fryeburg, Me	6.3	5.2	....
Germantown, Ohio.	5.6	5.4	3.1
Gettysburg, Pa	5.2	5.4	4.3
Glenwood, Tenn.	3.7	4.3	2.9
Gouverneur, N. Y.	7.2	6.7	4.9
Grand Rapids, Mich	4.4	4.9	3.0
Granville, Ohio	4.4	4.5	3.4
Green Springs, Ala	3.5	4.2	2.8
Harrisburg, Pa	5.7	5.4	4.3
Hillsboro, Ohio	3.3	3.3	....

NAME OF STATION.	7 A. M.	2 P. M.	9 P. M.
Hirum, Ohio.	3.5	3.9	2.6
Jacksonville, Fa	1.6	2.0	1.0
Jackson, Miss.	3.5	3.0	1.8
Jackson, Ohio, (Gilmor)	3.4	3.6	3.4
Key West, Fa	5.4	5.3	....
Knox Hill, Fa	3.5	4.2	3.0
Lewisburg, Va	3.8	3.7	3.7
Lima, Pa	5.4	5.9	3.9
Lodi, N. Y	7.1	7.0	6.1
Londonderry, N. H	5.3	4.6	5.3
Manchester, Ill.	3.5	3.9	4.2
Manchester, N. H	6.2	7.4	6.5
Meadville, Pa	6.2	6.6	5.1
Mendon, Mass	4.9	5.8	4.0
Millersburg, Ky	4.5	4.1	2.1
Milton, Ind	3.8	3.3	2.7
Milwaukie, Wis. (Winkler)	3.9	3.7	3.8
Monroe, Mich. (Whelpley)	4.7	5.5	5.7
Montreal, Ca.	6.2	6.9	7.1
Morrisville, Pa	3.7	3.7	3.2
Moss Grove, Pa	6.2	6.5	5.4
Muscatine, Iowa.	4.5	4.3	4.2
Nantucket, Mass	4.7	4.1	5.1
New Bedford, Mass	6.1	5.7	6.7
Newburyport, Mass	7.5	7.0	9.1
New Harmony, Ind	3.1	4.1	3.5
New London, Ct	4.2	4.8	5.0
New Orleans, La	4.2	3.3	2.1
New Wied, Texas	5.0	6.1	7.3
Norristown, Pa	4.2	5.4	4.1
North Attleboro, Mass	5.1	5.1	4.9
Oberlin, Ohio	4.7	4.4	3.0
Ottawa, Ill	4.0	3.6	3.9
Oxford, Miss	2.9	3.3	2.3
Penn Yan, N. Y	7.1	6.8	7.3
Perry, Me	6.7	7,3	5.8
Philadelphia, Pa.	6.4	5.7	4.1
Pittsburg, (Marine Hos-	7.4	6.5	4.3
pital,) Pa.(3)	....	....	....
Platteville, Wis	3.8	4.2	3.0
Pocopson, Pa	5.9	6.2	4.3
Pomfret, Conn	4.8	5.5	6.1
Portland, Me.	7.1	6.9	7.4
Pottsville, Pa.	6.5	6.0	4.5
Poultney, Iowa	4.0	5.9	4.6
Princeton, Mass	6.1	6.3	5.7
Providence, R. I.	5.3	4.9	5.3
Quasqueton, Iowa	4.8	4.5	4.1
Red River Settlement	7.3	5.3	5.4
Richmond, Mass	7.0	7.2	7.3
Richmond, Ohio	6.3	4.3	3.1
Sag Harbor, N. Y	4.5	5.3	6.1
St. James, Mich	6.2	6.0	6.7
St. Louis, Mo	3.6	3.6	3.4
St. Martin, Ca. E	4.9	6.1	6.8

* For changes in hours of observation, &c., see tables of Thermometer for this month.

(1) Monthly mean 6.4.

(2) Monthly mean 7.2.

(3) Obs. taken at sunrise, 9 A. M., 3 P. M., and 9 P. M.—Mean at 9 A. M., 7.1.

NAME OF STATION.	7 A. M.	2 P. M.	9 P. M.	NAME OF STATION.	7 A. M.	2 P. M.	9 P. M.
Saugatuck, Mich	4.8	5.8	4.1	Superior, Wis	3.3	3.3	3.1
Savannah, Ohio	5.7	5.1	2.5	Unionville, Ohio	6.2	6.3	6.8
Savannah, Ga	3.1	3,7	2·3	Upper Alton, Ill	2.6	2.9	1.7
Saybrook, Conn	4.8	4.8	4.9	Urbana, Ohio	4.9	4.0	3.0
Schellman Hall, Md	4.1	4.2	3.0	Wampsville, N. Y	8.1	7.7	5.8
Smithsonian Inst	5.0	4.0	3.3	Warrington, Fa	2.7	3·2	3.8
Smithville, N. Y	6.8	7.2	6.1	West Haverford, Pa	4.2	6.3	....
Southwick, Mass. (1)	5.9	6.4	6.0	Whitemarsh Island, Ga.	2.9	3.4	1.9
Sparta, Ga	2.9	3.4	1.6	Williamstown, Mass	7.7	6.8	7.1
Spencertown, N. Y	5.3	5.5	5.5	Winchester, Va	3.8	4.4	3.5
Springdale, Ky.	2.9	3.5	3.2	Woodward High School	3.9	3.6	2.7
Springfield, Mass	5.3	5.7	5.2	Worcester, Mass	4.3	4.9	4.5
Steuben, Me	7.7	7.5	6.9	Zanesville, Ohio	6.5	5.4	5.7

(1) Record incomplete.

NAME OF STATION.	7 A. M.	2 P. M.	9 P. M.	NAME OF STATION.	7 A. M.	2 P. M.	9 P. M.
Alexandria, Va	7.2	5.5	5.2	Hillsboro, Ohio	6.5	5.1	....
All Saints, S. C	5.1	4.6	4.0	Jackson, Miss	3.2	3.0	2.4
Amherst, Mass	5.8	5.0	6.2	Jackson, O., (Gilman)	6.6	5.2	6.1
Angelica, N. Y.	6.7	7.0	6.0	Jacksonville, Fa	5.8	6.1	3.7
Ann Arbor, Mich	.....	.....	.....	Key West, Fa	5.6	5.2	....
(Woodruff)	6.1	5.9	5.9	Knox Hill, Fa	6.9	7.4	5.4
Ann Arbor, " (Winchell.)	5.8	6.3	5.5	Lewisburg, Va	7.1	6.4	5.3
Annapolis, Md	6.7	5.4	4.8	Lima, Pa	6.7	6.2	5.2
Ashland, Va.	6.0	4.5	4.0	Lodi, N. Y.	7.6	7.3	7.0
Athens, Ill.	5.6	5.8	5.0	Londonderry, N. H.	5.4	5.1	5.2
Auburn, Ala	6.9	7.2	5.5	Manchester, Ill.	5.5	5.9	3.9
Augusta, Ill.	5.3	5.6	4.2	Manchester, N. H	5.7	5.9	5.1
Austin, Texas.	6.9	7.1	4.0	Meadville, Pa	7.8	6.9	5.6
Baldwinsville, N. Y.	7.9	7.5	5.6	Mendon, Mass	5.4	5.9	5.6
Battle Creek, Mich.	5.2	5.2	4.9	Millersburg, Ky	8.0	7.4	7.9
Bedford, Pa	6.4	5.2	4.6	Milton, Ind	6.0	5.2	4.7
Beloit, Wis.	3.7	2.3	2.5	Milwaukie, Wis. (Winkler)	4.7	4.7	3.5
Beverly, N. Y	6.3	6.0	6.4	Monroe, Mich	6.0	5.8	5.4
Blackwell's Isl'd, N. Y.	5.9	5.0	5.3	Montreal, Ca	4.8	6.2	5.5
Bladensburg, Md	7.1	5.6	4.8	Morrisville, Pa	5.0	4.6	4.1
Bloomfield, N. J	5.8	4.3	5.0	Moss Grove, Pa	6.7	6.5	6.3
Brandon, Vt	6.3	5.5	4.8	Muscatine, Iowa	5.6	4.7	3.7
Burlington, Vt	3.5	3.3	3.3	Nantucket, Mass.	5.4	5.5	6.0
Burlington, N. J.	8.2	6.6	7.6	New Bedford, Mass	6.6	6.3	6.8
Camden, Ark	7.1	5.7	4.4	Newburyport, Mass	8.4	7.5	8.8
Camden, S. C	6.5	5.8	5.6	New Harmony, Ind	6.2	5.5	5.5
Canton, N. Y	6.3	7.1	5·2	New London, Ct	5.1	5.0	5.0
Carmel, Me	6.3	6.2	4.9	New Orleans, La	8.0	7.2	5.1
Castleton, Vt	6.7	6.1	4.5	New Wied, Texas	4.7	5.9	7.5
Cedar Keys, Fa	4.7	4.8	2.9	Norristown, Pa.	6.8	7.0	5.0
Chapel Hill, N. C	3.4	3.0	5.1	North Attleboro, Mass.	6.0	5.3	5.5
Chestertown, Md	5.8	5.9	4.1	Oberlin, Ohio	6.4	6.0	4.9
Cleveland, Ohio	7.6	7.3	5.2	Ogdensdurg. N. Y.	5.7	5.6	....
Concord, N. H.	5.2	5.7	3.9	Ottawa, Ill	4.3	5.0	4.8
Cooper, Mich	5.1	4.6	3.5	Oxford, Miss.	6.7	6.7	5.1
Craftsbury, Vt.	7.3	6.8	6.5	Penn Yan, N. Y	7.5	7.1	7.4
Crichton's Store, Va	7.0	6.4	6.4	Perry, Me	6.5	6.3	6.4
Danville, Ky	4.8	4.2	3.8	Philadelphia, Pa	7.7	6.3	5.7
Deaf & Dumb Inst, N. Y.	6.5	6.5	5.4	Pittsburg, Pa. (Marine	....	....	....
Detroit, Mich	6.4	6.2	4.7	Hospital.) ..(1)	7.2	6.1	5.7
Delaware College	6.6	5.2	4.0	Platteville, Wis	5.2	4.5	5.2
Dubuque, Iowa	5.9	4.8	3.9	Pocopson, Pa	7.6	7.1	5.6
Easton, Pa	4.6	6.2	5.0	Pomfret, Conn	6.3	5.7	5.4
Exeter, N. H	6.2	6.1	4.6	Portland, Me	6.4	6.2	4.8
Ft. Madison, Iowa	4.9	4.6	4.4	Pottsville, Pa.	7.0	6.4	6.5
Frederick, Md	6.5	5.8	5.2	Poultney, Iowa	5.3	5.1	5.4
Fryeburg, Me	5.3	6.0	....	Princeton, Mass	5.7	6.3	5.3
Germantown, Ohio	7.3	6.2	6.3	Quasqueton, Iowa	6.1	4.5	4.2
Gettysburg, Pa.	6.5	5.1	5.4	Red River Settlement	6.7	5.9	5.4
Glenwood, Tenn.	5.7	7.3	4.9	Richmond, Mass	7.2	7.5	6.2
Gouverneur, N. Y	4.7	4.8	3.1	St. Augustine, Pa	4.0	4.1	3.0
Grand Rapids, Mich	5.0	5.9	4.4	St. James, Mich	6.2	6.8	6.9
Granville, Ohio	6.6	6.3	5.7	St. Louis, Mo	6.6	4.7	4.4
Green Springs, Ala	7.4	7.2	8.0	St. Martin, Ca., E	4.4	6.0	4.5
Hamlin University, Min.	5.1	5.6	6.6	Saugatuck, Mich	6.4	6.4	5.3
Harrisburg, Pa	6.7	5.5	6.2	Savannah, Ohio	7.0	6.9	5.6

* For changes in hours of observation, &c., see tables of Thermometer for this month.
(1) Obs. taken at sunrise, 9 A. M., 3 and 9 P. M.— mean at 9 A. M., 6.8.

NAME OF STATION.	7 A. M.	2 P. M.	9 P. M.	NAME OF STATION.	7 A. M.	2 P. M.	9 P. M.
Savannah, Ga.	6.8	6.7	4.3	Unionville, Ohio	7.4	7.2	6.3
Saybrook, Ct.	5.4	5.9	5.9	Upper Alton, Ill.	5.0	4.5	3.4
Schellman Hall, Md	6.1	5.6	4.3	Wampsville, N. Y.	7.8	7.4	6.4
Seneca Coll. Inst. Ovid,	....	....	....	Warrington, Fa	6.1	6.8	5.9
N. Y.	7.6	7.0	6.7	West Haverford, Pa	6.4	6.4	....
Smithsonian Inst	7.7	4.8	4.9	Whitemarsh Island, Ga.	5.2	6.2	4.7
Smithville, N. Y	6.4	6.4	5.9	Williamstown, Mass	7.5	6.6	6.2
Southwick, Mass	6.1	7.5	5.7	Winchester, Va	5.8	5.7	5.1
Sparta, Ga	6.1	7.1	5.8	Windham, Me(1)	4.1	4.7	4.5
Spencertown, N. Y	5.2	5.3	6.0	Woodward High School,	....	....	....
Springdale, Ky	6.1	5.4	5.8	(Cincinnatti, Ohio)	4.1	5.9	5.3
Springfield, Mass.	5.8	5.3	5.1	Worcester, Mass	4.6	4.9	5.4
Steuben, Me.	6.0	6.5	6.6	Zanesville, Ohio	8.0	6.0	6.4
Superior, Wis	4.0	4.1	3.3		....	....	....

NAME OF STATION.	7 A. M.	2 P. M.	9 P. M.
Alexandria, Va	5.0	5.1	5.3
All Saints, S. C	5.7	6.0	5.3
Amherst, Mass	5.6	6.3	5.8
Angelica, N. Y	7.1	7.1	5.5
Annapolis, Md	6.0	5.8	3.9
Ann Arbor, Mich.	....	....	....
(Winchell)	6.3	6.3	7.2
Ann Arbor, " (Woodruff)	6.6	6.4	6.4
Ashland, Va	6.3	5.9	5.0
Athens, Ill	5.4	5.6	5.1
Augusta, Ill	4.7	4.8	5.2
Austin, Texas	5.7	5.7	5.0
Baldwinsville, N. Y	6.7	7.4	6.7
Battle Creek, Mich	7.8	6.7	6.6
Beloit, Wis.	4.4	3.3	4.2
Beverly, N. Y	5.1	6.1	5.1
Bladensburg, Md	7.2	6.2	6.7
Bloomfield, N. J.	3.8	4.7	3.9
Boston, Mass..(1)	....	....	....
Brandon, Vt	6.3	6.4	4.9
Burlington, Vt	4.0	4.3	3.5
Burlington, N. J	8.8	6.5	7.5
Camden, S. C	6.3	6.5	5.9
Canton, N. Y	7.0	7.1	6.8
Carmel, Me	3.8	5.6	5.2
Cedar Keys, Fa	5.5	5.0	3.8
Chapel Hill, N. C	4.1	4.1	4.3
Chestertown, Md	5.6	4.7	4.5
Cincinnati, Ohio.(2)	....	....	....
Cleveland, Ohio	7.7	7.1	6.6
Concord, N. H	4.8	4.3	3.0
Craftsbury, Vt	6.6	6.4	5.5
Crichton's Store	5.2	5.0	4.1
Danville, Ky	6.3	5.6	5.8
Deaf and Dumb Inst., N.Y.	5.6	5.9	5.6
Detroit, Mich	7.4	7.5	5.4
Dickinsons College, Pa	4.4	5.6	5.1
Dubuque, Iowa	5.8	5.6	4.9
Easton, Pa.	4.4	6.5	4.5
Exeter, N. H	4.8	5.2	3.7
Fort Madison, Iowa	8.5	8.5	8.2
Frederick, Md	6.2	6.3	5.9
Fryeburg, Me	5.2	4.6	....
Germantown, Ohio	7.1	6.8	6.2
Gettysburg, Pa	4.5	5.4	5.0
Glenwood, Tenn.	6.5	6.0	4.9
Gouverneur, N. Y	8.1	7.9	6.6
Grand Rapids, Mich	6.4	5.8	6.0
Granville, Ohio	6.3	5.6	5.2
Hamline University, Min.	4.8	4.1	4.4
Harrisburg, Pa	5.3	5.9	5.8
Hillsboro, Ohio.	5.7	5.1	....
Jackson, Miss	2.4	2.4	1.8
Jacksonville, Fa	5.6	4.6	3.5
Key West, Fa	5.5	4.9	....
Lacquiparle, Min	4.3	4.0	2.5
Lewisburg, Va	7.7	6.1	6.1
Lima, Pa	5.2	6.4	5.6
Lodi, N. Y	7.3	7.5	7.2
Londonderry, N. H	5.2	4.5	4.2
Manchester, Ill	6.1	5.8	4.9
Manchester, N. H	5.5	4.4	4.9
Meadville, Pa	7.6	7.9	5.9
Mendon, Mass	4.9	5.2	4.2
Millersburg, Ky	5.3	5.5	4.6
Milton, Ind	6.1	5.4	5.9
Milwaukie, Wis.,	....	....	....
(Winkler)	6.1	6.1	4.3
Monroe, Mich, (Whelpley)	6.5	6.0	5.2
Montreal, Canada,	6.4	6.9	5.9
Morrisville, Pa.	3.5	4.4	5.8
Moss Grove, Pa	7.1	7.2	6.1
Muscatine, Iowa	5.5	5.8	6.2
Nantucket, Mass	6.1	5.1	5.3
Nazareth, Pa	3.9	6.4	5.9
New Bedford, Mass.	6.7	6.0	6.1
Newburyport, Mass.	7.8	7.1	8.9
New Harmony, Ind	6.5	5.1	5.2
New London, Ct	5.2	4.8	3.8
New Orleans, La	7.8	7.1	5.0
New Wied, Texas	5.0	5.8	6.6
Norristown, Pa	5.4	6.4	6.0
North Attleboro, Mass	4.6	5.1	3.2
Oberlin, Ohio	7.3	6.0	4.0
Ottawa, Ill	5.6	4.2	4.1
Oxford, Miss	5.1	4.4	2.9
Penn Yan, N. Y	7.6	7.6	7.2
Perry, Me.	4.3	5.3	4.1
Philadelphia, Pa	7.0	7.2	6.0
Pittsburg, Pa., (Marine	....	....	....
Hospital,)..(3)	7.8	6.9	6.2
Platteville, Wis	6.5	6.6	5.1
Pocopson, Pa	6.1	6.9	6.0
Pomfret, Conn.	5.6	5.0	5.0
Portland, Me	5.7	7.1	4.7
Pottsville, Pa	6.1	6.5	5.6
Princeton, Mass	7.0	6.0	4.8
Providence, R. I	4.9	5.0	4.2
Poultney, Iowa	6.9	6.6	5.5
Quasqueton, Iowa	6.9	5.9	4.7
Red River Settllement	5.1	5.0	3.4
Richmond, Mass	5.9	5.6	5.9
Sag Harbor, N. Y	5.2	5.8	4.8
St. James, Mich	8.4	7.6	8.4
St. Louis, Mo	5.5	4.7	4.6
St. Martin, Ca., E	5.4	5.9	6.8
Saugatuck, Mich	8.0	8.0	7.9
Savannah, Ohio	6.3	6.6	6.0
Savannah, Ga.	6.5	6.9	6.5
Saybrook, Ct	5.1	5.1	4.3
Schellman Hall, Md	5.9	6.0	4.6
Seneca Coll. Inst. Ovid,	....	....	....
N. Y	6.3	7.2	6.1
Smithsonian Inst	6.2	4.6	5.5

* For changes in hours of observation see tables of Thermometer for this month.

(1) Mean for the month, 3.5.

(2) Mean for the month, 5.5.

(3) Obs. taken at sunrise, 9 A. M., 3 P. M. and 9 P. M.—mean at 9 A. M. 8. 3.

NAME OF STATION.	7 A. M.	2 P. M.	9 P. M.	NAME OF STATION.	7 A. M.	2 P. M.	9 P. M.
Smithville, N. Y.	7.4	7.4	7.5	Warrington, Fa.	6.5	6.6	6.2
Southwick, Mass.	4.7	6.7	7.5	Westfield, Mass.	5.0	4.7	4.8
Sparta, Ga.	6.4	6.8	5.4	West Haverford, Pa.	4.6	6.1	....
Spencertown, N. Y.	4.6	4.9	4.0	Whitemarsh Isl'd, Ga.	5.3	7.0	6.0
Springdale, Ky	5.8	6.0	5.7	Williamstown, Mass.	6.6	6.6	5.6
Springfield, Mass.	4.3	5.1	4.1	Winchester, Va.	5.6	5.7	5.6
Steuben, Me.	5.2	5.9	5.1	Windham, Me.	3.0	3.6	3.2
Superior, Wis.	6.7	5.7	5.0	Woodward High School,	....	....	....
Unionville, Ohio.	7.4	7.8	7.0	Cincinnnati, Ohio.	8.2	6.7	5.7
Upper Alton, Ill.	4.1	4.0	2.3	Worcester, Mass.	5.0	5.5	5.4
Urbana, Ohio	7.2	5.9	6.3	Zanesville, Ohio.	8.2	6.6	7.3
Wampsville, N. Y.	6.8	7.5	6.3		....	....	....

NAME OF STATION.	JANUARY.	FEB.	MARCH.	APRIL.	MAY.	JUNE.	JULY.	AUGUST.	SEPT.	OCTOBER.	NOV.	DEC.	TOTAL FOR THE YEAR.
Aiken, S. C.	2.084	2.193	2.715	1.308	5.600	4.789	6.177	5.621	2.147	0.240	1.810	6.133	40.817
Alexandria, Va.	2.246	1.770	1.526	1.902	1.412	6.333	4.272	3.896	2.671	2.990	0.820	3.480	33.318
All Saints, S. C.	1.090	1.790	3.470	0.540	0.970	7.110	3.600	1.910	1.280	1.150	1.330	6.730	30.970
Amherst, Mass.	5.061	2.702	1.078	3.845	1.492	5.191	6.102	2.553	0.547	10.078	4.123	5.409	48.181
Ann Arbor, Mich.:													
Woodruff	3.380	2.300	2.360	5.910	0.963	7.089	8.740	1.504	5.265	2.166	5.398	3.948	49.023
Winchell	2.450	1.880	2.182	5.668	0.914	6.574	7.668	1.488	4.424	2.266	4.432	3.391	43.337
Arcola, Ohio	5.375	...	...	...	...	...	5.125	...	...	...	...	...	...
Ashland, Va.	1.304	...	2.859	1.489	1.671	5.300	3.669	5.183	3.614	1.319	2.068	3.567	...
Athens, Ill.	9.370	0.850	2.560	1.600	4.830	3.400	5.140	4.400	2.300	2.530	2.590	3.950	43.520
Auburn, Ala.	0.250	0.990	1.850	2.280	1.720	1.890	...	5.860	1.640	2.470	5.380	9.340	...
Baldwin Institute, Ohio	...	...	2.526	...	...	...	...	...	...	...	...	...	...
Battle Creek, Mich.	3.240	1.420	1.350	8.350	1.080	5.400	6.460	0.690	7.270	1.870	4.180	1.630	42.940
Bedford, Pa. (1)	?	0.960	1.818	1.624	1.933	6.060	3.932	2.870	...	1.065	6.720	...	...
Beloit, Wis.	...	...	...	...	1.745	4.607	3.090	1.380	2.310	2.800	1.605	1.950	...
Blackwell's Island, N. Y.	...	...	...	...	...	...	...	...	...	...	2.597	...	...
Bloomfield, N. J.	3.193	3.366	1.642	1.420	2.649	4.450	3.930	1.956	1.064	3.826	2.512	2.875	32.883
Burlington, Vt.	1.950	0.980	0.720	2.220	0.340	6.470	8.430	2.940	4.500	5.970	1.280	2.500	38.300
Camden, S. C.	1.320	1.060	4.400	0.850	5.220	7.490	2.740	7.290	5.220	0.720 *	1.640	4.650	42.600
Camden, Ark.	...	...	...	...	...	...	...	...	...	3.300	4.980	...	...
Canton, N. Y. (1)	?	?	?	2.500	?	?	?	?	?	?	?	?	...
Cedar Keys, Fa.	1.140	1.437	2.937	0.375	0.625	3.250	7.500	5.125	1.937	5.187	2.437	2.562	34.512
Ceresco, Wis.	?	?	...	2.700	4.950	...	...	...	...	...	...	...	...
Charleston, S. C.	0.750	0.810	4.410	0.880	5.390	3.690	4.420	2.690	3.280	0.860	0.970	6.700	34.850
Chestertown, Md.	...	...	...	...	...	4.100	3.030	9.440	...	4.890	1.650	4.140	...
Cincinnati, Ohio, (Lea)	...	...	...	...	...	...	...	...	...	...	...	3.280	...
Cleveland, Ohio	...	...	...	...	2.900	10.330	2.030	3.340	2.520	3.470	4.300	3.030	...
Columbus, Miss.	...	...	...	...	5.250	3.672	2.952	3.607	5.204	1.532	8 877	4.304	...
Concord, N. H. (1)	?	...	?	5.250	1.000	4.230	4.750	3.250	0.250	11.000	4.250	?	...
Cooper, Mich. (1)	?	?	?	?	3.125	17.125	12.500	6.687	14.125	8.000	...	...	...
Crichton's Store, Va.	2.300	...	5.650	0.350	1.750	5.400	4.700	7.200	0.550	3.100	3.000	4.900	...
Danville, Ky.	3.217	0.860	5.501	1.349	...	5.464	3.370	...	5.761	...	3.420	3.184	...

(1) Snow not reduced to water in all cases.

* First twelve days.

NAME OF STATION.	JANUARY	FEB.	MARCH.	APRIL.	MAY.	JUNE.	JULY.	AUGUST.	SEPT.	OCTOBER.	NOV.	DEC.	TOTAL FOR THE YEAR.
Deaf and Dumb Inst., N. Y.	4.770	5.120	2.330	2.860	4.900	5.830	5.460*	2.900	1.510	7.370	3.000	6.880	52.930
Detroit, Mich	4.535	1.550	4.009	7.390	2.056	11.657	15.010	1.962	5.698	4.391	9.213	3.720	71.191
Dubuque, Iowa	1.600	1.130	1.300	0.640	2.020	4.380	3.070	1.910	6.060	2.500	2.710	1.810	29.130
Flint, Mich	2.845	1.463	1.740	2.490	2.549	7.601	3.963	0.850	5.760	2.545	5.574	3.171	40.551
Fort Madison, Iowa, (1)	?	?	1.025	4.850	8.050	5.100	6.900	4.350	3.250	2.500	4.850	?	
Frederick City, Md	3.055	1.882	2.290	1.133	2.160	5.687	2.316	3.804	5.495	2.103	1.058	3.076	34.059
Friendship, Tenn			1.9375	4.850	4.700	6.500	8.110						
Fryeburg, Me. (1)	?	?	?	?	0.562	5.775	3.500	3.750	?	9.187	6.333	?	
Gallipolis, Ohio	1.544	1.699	2.381	1.543	3.755	5.380	4.076	4.103	5.040	1.300	2.540		
Gardiner, Me	7.169	1.700	1.103	4.570	1.932	5.984	2.416	3.096					
Germantown, Ohio	4.192	1.437	3.554	2.285	2.537	6.925	5.190	5.043	4.645	1.289	6.260	3.648	47.005
Gettysburg, Pa	3.572	2.410	2.040	2.085	1.690	7.716	3.790	10.885	5.915	2.220	1.268	4.030	47.621
Glenwood, Tenn	4.667	1.531	3.851	2.598	3.783	3.923	3.572	4.560	7.298	1.400	4.657	2.265	44.105
Gouverneur, N. Y	2.580	0.100	1.650	0.850	0.400	6.600	3.050	1.300	1.750	4.900	0.150	2.330	25.660
Grand Rapids, Mich	6.800	0.940	2.310	3.540	0.940	6.720	7.240	4.690	5.280	4.190	4.380	5.030?	52.060
Granville, Ohio	2.770	1.900	5.500	2.550	3.220	7.840	7.670	3.150	6.640	1.700	6.900	5.040	54.880
Green Springs, Ala	1.090	1.200	1.190	0.400	1.300	4.610	3.130	6.620		1.640	7.110		
Great Falls, N. H. (1)	4.345	6.320	?	5.475	1.130	4.470							
Hamline University, Min											1.373	1.247	
Hannibal, Mo. (1)	?	0.562	2.000	0.937	7.000								
Harrisburg, Pa. (1)	?	?	?	1.879	2.955	8.541	9.188	3.150	3.681	4.815	1.565	?	
Hillsborough, Ohio				1.150									
Hiram, Ohio									3.125	3.000			
Jackson, Miss										1.560	10.930	6.820	
Jackson, Ohio, (Gilmor)				1.340	2.968	4.100	4.310	6.510	5.890	1.030	2.952		
Jackson, Ohio, (Wood)				1.340									
Jacksonville, Fa	1.450	1.650	5.350	1.150	1.000	5.425	2.150	6.150	4.340	2.550	2.725	4.000	37.940
Janesville, Wis	3.400	1.250											
Key West, Fa	4.110	1.500	1.430	0.260	2.480	3.660	3.850	5.860	7.280	3.270	2.780	2.130	38.610
Knox Hill, Fa	2.9202	2.8315	1.7000	3.1689	2.9267	5.2327	8.4078	5.6059	2.3198	1.9607	7.6121		
Knoxville, Tenn	2.610	1.170	4.130	2.300	3.230	4.780							
Lac qui Parle, Min					?	5.000	?	10.500				?	

(1) Snow not reduced to water in all cases.

*Approximate.

NAME OF STATION.	JANUARY.	FEB.	MARCH.	APRIL.	MAY.	JUNE.	JULY.	AUGUST.	SEPT.	OCTOBER.	NOV.	DEC.	TOTAL FOR THE YEAR.
La Grange, Ga	0.210												
Lebanon, Tenn			2.980										
Lewisburg, Va							3.360						
Lima, Pa	2.410	2.360	1.660	2.300	4.300	9.320	4.060	2.480	3.920	5.350	2.010	6.410	46.580
Lodi, N. Y	2.4375	1.500	0.625	4.000	2.9375	5.8125	5.000	2.500	3.1875	4.3125	2.000	2.6875	37.000
Londonderry, N. H	6.950	4.950	0.400	3.450		2.800		1.800	0.030	7.400	3.100	5.400	
Lowville, N. Y	4.812	2.375	3.875	2.375	1.312	6.125	4.437*						
Madison, Ohio		2.885											
Madison, Wis					4.438								
Madrid, N. Y					?	4.050	3.700	3.220					
Manchester, Ill	13.526	1.562	2.910	3.969	6.431	2.000	3.521	2.065	2.765	2.551	2.820	3.922	48.042
Marietta, Ohio	1.708	1.090	2.080	1.805	4.085								
Millersburg, Ky	1.910	0.500	5.100	1.910	2.380	4.840				0.552	2.270	3.500	
Milwaukie, Wis. (Winkler)	1.800	1.200	1.750	2.300	1.450	3.680	3.560	3.090	6.880	2.110	1.850	2.610	32.280
Monroeville, Ala	0.250	1.625					6.000	6.625	5.687	?	6.250		
Montreal, Canada									3.860	5.370	5.770	4.130	
Morrisville, Pa	2.600	1.800	1.800	2.500	3.400	8.800	5.000	2.500	2.300	4.900	2.700	4.800	43.100
Moss Grove, Pa		3.300	1.700	2.950	3.000	7.300	3.650	5.700	3.600	3.100	4.800	3.750	
Mount Vernon, Ohio (1)	1.120	?	3.090	1.120									
Muscatine, Iowa (1)	1.500	?	2.520	2.560	1.940	4.750	2.350	3.510	1.840	2.810	2.080	?	
Nantucket, Mass	4.504	3.040	3.838	8.492	5.875	4.772	2.162	1.509	1.210	4.130	3.106	6.299	48.937
Newark, Ohio	2.000	?	2.500	?									
Newark, N. J	4.030	3.466	1.875	2.470	2.365	4.525	4.470	4.160	2.250	5.260	2.890	6.500	44.261
New Bedford, Mass.	4.200	2.330	1.730	3.770	3.230	1.630	4.260	1.260	0.550	4.145	4.290	5.050	36.445
Newburyport, Mass.	5.343	3.795	1.070	5.280	1.385	2.418	2.568	3.135	0.460	6.860	4.300	5.595	42.209
New Harmony, Ind	4.660	0.750	4.400	5.020	3.540	5.190	7.010	4.440	5.680	1.320	3.650	2.450	48.070
New Lisbon, Ohio	1.900	1.000	2.100	1.100	3.000								
New Orleans, La	0.850	1.720	0.900	1.800	2.090	5.900	3.877	4.160	5.110	2.055	5.230	5.275	38.967
New Wied, Texas	0.020	2.150	0.890	0.450	1.220	4.020	3.470	5.500	1.800	4.700	1.000	0.200	25.420
Norristown, Pa	2.600	3.100	1.850	1.970	4.370	7.550	7.580	1.340	5.230	4.940	2.180	5.430	48.140
North Attleboro, Mass	5.740	4.084	0.900	3.700	2.910	3.685	5.274	2.025	0.605	5.955	4.480	5.580	44.938
Norwalk, Ohio	2.640												

(1) Snow not reduced to water in all cases.

*Approximate.

NAME OF STATION.	JANUARY.	FEB.	MARCH.	APRIL.	MAY.	JUNE.	JULY.	AUGUST.	SEPT.	OCTOBER.	NOV.	DEC.	TOTAL FOR THE YEAR.
Oberlin, Ohio	2.450	1.360	1.440	4.930	2.050	10.650	3.590	6.720	4.520	3.550	4.310	2.490	48.060
Oxford, Miss.	2.837	2.444	1.985	3.835	1.697	5.641	4.013	5.440†	1.113‡	1.451	8.675	2.337	41.468
Penn Yan, N. Y.	2.000	1.560	0.670	2.850	3.020	5.960	4.560	1.700	1.340	3.500	1.280	1.520	29.960
Perry, Me.	2.500	?	1.300	4.600	5.600	8.500	2.700	2.800	1.650	10.500	3.700	4.000	......
Perrysburg, Ohio	2.000	1.375	2.500	4.687	3.000	8.125	9.812	2.437	7.500	2.250	7.000	4.500	55.186
Philadelphia, Pa.	2.601	2.480	1.979	2.148	3.033	8.008	6.594	3.237	4.129	3.416	2.022	5.006	44.653
Pittsburg, Pa., (Oakland Station)	0.820	1.640	3.300	2.380	2.195	7.725	8.823	5.396	4.728	1.365	5.347	2.193	45.912
Pocopson, Pa.	2.820	3.390	1.810	2.060	3.720	8.110	3.260	4.130	7.080	4.680	2.240	6.260	49.560
Pomfret, Conn.	5.790	4.100	0.870	2.610	1.990	3.220	4.570	2.040	0.220	7.190	3.930	6.150	42.680
Port Gibson, Miss.	......	......	......	......	......	......	......	3.950§	......	......	......	......	......
Pottsville, Pa.	3.326	3.131	?	2.567	2.377	9.461	8.432	2.160	4.762	3.988	2.198	5.786	......
Poultney, Iowa,*	2.237	0.860	2.016	1.050	2.133	3.265	2.875	4.705	4.845	2.014	2.618	2.370?	30.988
Princeton, Mass.	5.240	5.400	1.140	4.480	1.080	3.200	6.030	2.200	0.760	8.360	4.520	4.830	47.240
Providence, R. I.	6.450	4.050	0.850	2.500	2.550	1.950	3.250	2.020	0.250	5.330	3.750	6.100	39.050
Quasqueton, Iowa, (1)	1.005	?	?	?	1.294	2.740	2.480	......	......	1.250	?	......	......
Randolph, Pa.	4.400	3.200	2.850	3.050	3.350	9.900	......	......	3.650	......	......	......	......
Red River Settlement, Hudson's Bay Territory	......	......	......	......	......	5.125	15.125	13.500	6.000	......	......	......	......
Richmond, Mass. (1)	?	......	?	8.725	8.417	......	22.187?	10.000	3.000	19.375	?	?	......
Sacramento, Cal.	2.670	3.460	4.200	......	......	......	......	......	......	......	......	......	......
Sag Harbor, N. Y.	5.320	1.250	1.380	4.050	3.200	2.660	4.370	2.980	0.900	5.380	......	4.910	......
St. Augustine, Fa.	......	......	......	......	......	......	......	......	......	......	1.800	......	......
St. James, Mich.* (1)	1.250	......	......	0.750	0.500	?	?	0.875	1.875	1.375	?	......	......
St. Johnsbury, Vt. (1)	?	......	?	1.000	0.120	10.250	3.920	......	......	......	......	......	......
St. Louis, Mo.	4.660	0.700	2.890	2.650	7.460	4.270	5.170	6.530	3.890	3.890	5.160	3.100	50.370
St. Martin's, Canada East.	3.203	1.500	2.091	3.712	1.756	8.417	2.351	4.366	3.471	8.894	4.658	7.242	51.661
San Francisco, Cal.	......	......	4.310	......	......	......	......	......	......	......	......	......	......
Saugatuck, Mich.	......	......	?	?	?	?	......	?	4.025	4.580	3.507	3.512	......
Savannah, Ga.	1.261	1.176	2.771	2.211	5.887	4.944	2.645	5.582	2.251	1.982	2.224	5.630	38.594
Savannah, Ohio	......	......	......	3.000	4.483	16.600	8.616	3.475	6.791	3.433	6.082	3.940	......
Saybrook, Conn.	5.000	......	1.050	4.000	2.500	3.125	3.750	?	0.750	7.500	2.750	6.500	......

(1) Snow not reduced to water in all cases. * Amount not always given. † From Aug. 1 to Sept. 9. ‡ Since Sept. 9. § Since the 19th.

NAME OF STATION.	JANUARY.	FEB.	MARCH.	APRIL.	MAY.	JUNE.	JULY.	AUGUST.	SEPT.	OCTOBER.	NOV.	DEC.	TOTAL FOR THE YEAR.
Schellman Hall, Md. (1)	3.500	2.150?	4.950	1.550	3.000	7.450	4.200	4.000	10.500	4.500	1.750	6.250?	53.800?
Smithfield, Va. (1)	3.067	1.120?	6.055?	2.274	1.382	4.218	5.385	5.026	2.885	5.098	0.933	4.290	41.733?
Smithsonian Inst., D. C.	…	…	…	?	1.421	5.573	4.056	2.903	2.913	3.535	1.115	3.200	…
Southwick, Mass. (1)	?	2.810	1.890	3.070	2.580	4.020	7.790	2.460	1.380	11.810	3.960	6.080?	…
South Thomaston, Me.	…	…	…	…	…	25.000*	…	…	…	…	…	…	…
Sparta, Ga.	1.020	1.980	2.260	1.650	4.030	4.970	11.740	11.300	0.930	1.220	2.090	8.330	51.520
Spencertown, N. Y.	1.586	0.886	0.200	2.740	3.565	6.863	8.109	2.330	1.630	7.416	2.879	4.267	42.471
Springdale, Ky	4.840	1.210	5.070	…	3.750	8.100	2.550	4.440	3.650	1.850	5.160	3.180	…
Springfield, Mass	4.000	3.040	0.950	3.110	1.910	4.150	8.180	2.400	0.740	9.460	3.870	5.400	47.210
Steuben, Me. (1)	?	12.800*	?	9.900	5.500	6.300	1.800	5.100	1.900	17.000*	6.600	?	…
Thornbury, N. C.	2.416	0.830	4.030	1.953	…	…	…	…	…	…	…	…	…
Tuscaloosa, Ala.	0.901	1.734	2.236	…	…	…	…	…	…	…	…	…	…
Unionville, Ohio	…	…	…	3.344	6.875	6.750	…	…	3.594	7.781	6.750	4.765	…
Upper Alton, Ill.	4.827	0.060	2.500	0.924	3.947	3.081	3.009	1.944	2.818	3.200	3.030	3.050	32.390
Urbana, Ohio	…	1.660	3.400	…	6.720	10.780	6.170	1.230	8.760	3.180	…	3.860	…
Warrington, Fa.	0.187	1.250	1.562	0.750	0.844	3.437	11.125	10.562	2.187	3.500	6.062	8.500	49.966
Westfield, Mass	5.210	2.870	1.340	2.950	2.070	4.700	5.380	2.950	0.370	10.270	4.160	6.120	48.390
West Haverford, Pa.	2.072	3.330	1.680	?	2.725	7.927	5.282	2.069	?	?	?	?	…
Whitemarsh Island, Ga.	1.220	1.820	2.600	2.820	5.040	3.360	2.100	4.460	4.460	1.470	2.300	5.710	37.360
Williamstown, Mass.	2.520	1.430	0.210	5.640	2.850	5.480	6.750	3.557	1.915	12.931	4.592	6.629	54.504
Wood's Hole, Mass.	…	4.577	3.900	5.830	…	…	…	…	…	…	…	…	…
Woodward High School, O.	4.470	1.590	…	…	…	…	…	…	…	3.180	5.220	3.335	…
Worcester, Mass. (2)	8.010	4.480	2.300	5.390	1.640	4.190	9.400	4.060	?	7.970	5.850	6.900	…
Zanesville, Ohio	3.324	2.710	2.571	2.005	…	6.194	6.720	4.578	6.288	1.493	5.221	2.920	…

(1) Snow not reduced to water in all cases. (2) Observations in September incomplete. * Doubtful.

NAME OF STATION.	MEAN.				MAXIMA.		MINIMA.		RANGE FOR THE YEAR.
	7 A. M.	2 P. M.	9 P. M.	MEAN.	DATE.	HEIGHT.	DATE.	HEIGHT.	
Alexandria, Va	30.04	30.00	30.03	30.02	Jan. 8	30.80	Dec. 9	29.15	1.65
All Saints, S. C	30.02	30.00	30.01	30.01	Dec. 27	30.57	Mar. 31	29.23	1.34
Amherst, Mass	29.76	29.72	29.74	29.74	Jan. 6	30.66	Dec. 9	28.64	2.02
Ann Arbor, Mich., (Winchell)	29.05	29.02	29.04	29.04	Jan. 6	29.81	Dec. 9	27.96	
Beloit, Wis	29.19	29.17	29.18	29.18	Jan. 7	29.81	Dec. 9	28.29	1.52
Bloomfield, N. J	29.85	29.82	29.85	29.84	Jan. 8	30.65	Dec. 9	28.89	1.76
Burlington, Vt	29.63	29.61	29.62	29.62	Jan. 5	30.49	Dec. 9	28.56	1.93
Burlington, N. J. (1)	30.00	29.98	29.99	29.99	Jan. 8	30.78	Dec. 9	29.10	1.68
Camden, S. C	29.92	29.88	29.89	29.90	Jan. 8	30.49	De. 9, Ja. 28	29.33	1.16
Canton, N. Y	29.41	29.38	29.40	29.40	Jan. 8	30.27	Dec. 10	28.33	1.94
Carmel, Me	29.61	29.57	29.59	29.59	Jan. 6	30.65	April 1	28.39	2.26
Cedar Keys, Fa	29.92	29.91	29.91	29.91	De. 12, Ja 9	30.30	Jan. 28	29.52	.78
Chapel Hill, N. C	29.47	29.44	29.45	29.45	Jan. 8	30.07	Jan. 21	28.88	1.19
Charleston, S. C. (2)	....	....	....	....	Jan.	30.71	Oct.	29.32	1.39
Concord, N. H	29.63	29.60	29.61	29.61	Jan. 6	30.60	Dec. 9	28.53	2.07
Detroit, Mich	29.51	29.49	29.50	29.50	Jan. 8	30.30	Dec. 9	28.41	1.89
Dubuque, Iowa	29.33	29.28	29.30	29.30	Dec. 27	29.94	Dec. 8	28.25	1.69
Frederick, Md	29.62	29.58	29.60	29.60	Jan. 8	30.46	Dec. 9	28.61	1.85
Germantown, Ohio	29.25	29.21	29.23	29.23	Jan. 8	29.97	Jan. 21	28.25	1.72
Gettysburg, Pa	29.46	29.42	29.44	29.44	Jan. 8	30.20	Dec. 9	28.47	1.73
Glenwood, Tenn	29.59	29.54	29.57	29.57	Jan. 7	30.23	Jan. 21	28.77	1.46
Granville, Ohio	28.99	28.96	28.97	28.97	Jan. 8	29.72	Jan. 21	28.05	1.67
Harrisburg, Pa	29.76	29.73	29.75	29.75	Jan. 6	30.40	Dec. 9	28.83	1.57
Hillsborough, Ohio	28.89	28.86	28.88	28.88	Jan. 8	29.57	Jan. 21	28.14	1.37
Jacksonville, Fa	30.14	30.09	30.12	30.12	Jan. 8	30.60	Mar. 12	29.45	1.15
Key West, Fa	29.72	29.75	....	29.73	Jan. 7	30.05	Mar. 31	29.42	.63
Knox Hill, Fa	29.84	29.79	29.82	29.82	March 1	30.24	Jan. 28	29.31	.93
Lima, Pa	29.87	29.83	29.85	29.85	Jan. 8	30.62	Jan. 26	29.04	1.58
Manchester, N. H	29.95	29.93	29.94	29.94	Jan. 6	30.91	Dec. 9	28.91	2.00
Milton, Ind	29.06	29.03	29.06	29.05	Jan. 7	29.67	Jan. 21	28.07	1.60
Milwaukie, Wis. (Winkler)	29.29	29.27	29.28	29.28	Jan. 7	29.97	Dec. 9	28.25	1.72
Morrisville, Pa. (1)	30.12	30.10	30.11	30.11	Jan. 8	30.90	Dec. 9	29.20	1.70

(1) Not corrected for temperature.

(2) Mean for the year 29.99.

Name of Station.	Mean.				Maxima.		Minima.		Range for the year.
	7 A. M.	2 P. M.	9 P. M.	Mean.	Date.	Height.	Date.	Height.	
Muscatine, Iowa	29.50	29.46	29.48	29.48	Dec. 24	30.04	Dec. 8	28.56	1.48
Nantucket, Mass.	29.99	29.96	29.98	29.98	Jan. 6	30.86	Dec. 9	29.11	1.75
Newark, N. J. (1)	30.00	......	29.99	29.99	Jan. 6	30.79	Jan. 23	29.23	1.56
New Bedford, Mass	29.92	29.88	29.91	29.90	Jan. 6	30.75	April 1	28.87	1.88
Newburyport, Mass	29.97	29.94	29.96	29.96	Jan. 6	31.07	Dec. 9	28.92	2.15
New Orleans, La	30.23	30.20	30.22	30.22	Jan. 8	30.86	Jan. 21	29.85	1.01
New Wied, Texas (2)	28.86	28.78	28.81	28.82	Jan. 30	29.60	Jan. 22	28.38	1.22
Norristown, Pa	29.93	29.90	29.92	29.92	Jan. 8	30.71	{ Nov. 28 / Jan. 26 }	29.12	1.59
North Attleboro, Mass	29.90	29.86	29.89	29.88	Jan. 6	30.71	Dec. 9	28.76	1.95
Oberlin, Ohio	29.11	29.08	29.11	29.10	Jan. 8	29.89	Dec. 9	28.02	1.87
Oxford, Miss	29.58	29.54	29.55	29.56	Dec. 26	30.41	Jan. 21	29.04	1.37
Penn Yan, N. Y. (3)	29.22	29.23	29.23	29.23	Jan. 8	30.00	Dec. 9	28.18	1.82
Perrysburg, Ohio (3)	29.38	29.35	29.36	29.36	Jan. 8	30.16	Dec. 9	28.34	1.82
Philadelphia, Pa	29.89	29.85	29.87	29.87	Jan. 8	30.61	Dec. 9	28.98	1.63
Pittsburg, Pa., (Oakland Station)	29.81	29.77	29.78	29.79	Jan. 8	29.79	Dec. 9	27.94	1.85
Pittsburg, Pa., (Marine Hospital) (4)	28.97	28.95	28.96	28.96	Jan. 8	29.80	Ja. 21, De. 9	28.05	1.75
Pomfret, Conn	28.98	28.96	28.97	28.97	Jan. 6	29.70	Dec. 9	27.95	1.75
Pottsville, Pa	29.25	29.20	29.23	29.23	Jan. 8	30.05	Jan. 22	28.12	1.93
Princeton, Mass	28.82	28.80	28.81	28.81	Jan. 6	29.68	Dec. 9	27.76	1.92
Sag Harbor, N. Y. (5)	29.68	29.68	29.69	29.68	Jan. 6	30.58	Dec. 9	28.68	1.90
Savannah, Ga	30.09	30.04	30.07	30.07	Jan. 8	30.61	Mar. 31	29.40	1.21
Smithsonian Institution, Washington City	30.06	30.01	30.03	30.03	Jan. 8	30.81	Dec. 9	29.07	1.74
Southwick, Mass	29.97	29.92	......	......	Jan. 6	30.83	Jan. 26	29.06	1.77
Sparta, Ga	29.36	29.33	29.35	29.35	Jan. 8	29.92	Jan. 28	28.90	1.02
Spencertown, N. Y	29.29	29.26	29.28	29.28	Jan. 8	30.08	Jan. 26	28.25	1.83
Steuben, Me	30.02	29.99	30.00	30.00	Jan. 5	30.89	Dec. 26	29.19	1.70
The Rock, Ga. (3)	......	......	......	29.51	April	30.15	Jan.	29.10	1.05
Upper Alton, Ill	29.48	29.44	29.46	29.46	Jan. 7	30.14	Mar. 12	28.33	1.81
Warrington, Fa. (3)	30.04	30.05	30.01	30.03	Mr. 1, De. 12	30.48	Jan. 28	29.57	.91
Williamstown, Mass. (3)	29.26	29.22	29.26	29.25	Jan. 6	30.09	Dec. 9	28.20	1.89
Worcester, Mass	29.46	29.43	29.45	29.45	Dec. 13	30.24	Nov. 29	28.44	1.80

(1) Obs. at 7 A. M. and 6 P. M., and not corrected for temperature. (2) Not corrected for temperature. (3) For hours of obs., see monthly tables. (4) Obs. taken at sunrise, 9 A. M., 3 P. M., and 9 P. M.; Mean at 9 A. M., 28.95. (5) Year complete except November.

NAME OF STATION.	MEAN.				MAXIMA.				MINIMA.				RANGE FOR THE YEAR.
					WARMEST DAY.		HIGHEST DEGREE.		COLDEST DAY.		LOWEST DEGREE.		
	7 A. M.	2 P. M.	9 P. M.	MEAN.	Date.	Mean temp.	Date.	Temp	Date.	Mean temp.	Date.	Temp	
Alexandria, Va	49.41	60.33	52.52	54.09	June 29	87.33	July 19	95	Feb. 10	14.00	Feb. 7	6	89
All Saints, S. C	59.60	69.32	62.58	63.83	Aug. 11	85.33	Aug. 9	92	Feb. 27	30.33	March 2	25	67
Amherst, Mass	41.93	52.80	44.73	46.49	June 30	84.67	{ June 30, July 19 }	92	Feb. 6	-8.00	Feb. 7	-16	108
Angelica, N. Y	39.97	51.48	42.21	44.55	July 19	81.33	June 30	93	Feb. 6	-18.67	Feb. 6	-28	121
Ann Arbor, Mich., (Wood.)	40.99	53.34	45.20	46.51	July 16	82.00	{ June 28, July 16 }	91	Feb. 6	-5.67	Feb. 24	-24	115
Ann Arbor, Mich. (Winch.)	42.14	52.32	44.06	46.17	July 16	81.40	June 28	90	Feb. 6	-6.07	Feb. 6	-14	105
Athens, Ill	47.01	60.90	49.94	52.62	July 18	89.33	July 19	101	Dec. 25	-7.00	Dec. 26	-20	121
Augusta, Ill	44.22	58.67	48.91	50.60	Aug. 3	85.33	July 19	95	Dec. 26	-6.00	Dec. 30	-16	111
Austin, Texas	57.29	76.18	64.04	65.84	Aug. 5	87.33	March 8	98	Dec. 25	17.33	Dec. 26	7	91
Baldwinsville, N. Y	41.28	49.55	44.67	45.17	June 30	81.00	June 30	89	Feb. 6	-16.67	Feb. 6	-22	111
Battle Creek, Mich	43.61	56.63	45.07	48.44	July 16	87.33	July 16	96	Feb. 6	-2.00	Feb. 6	-9	105
Beloit, Wis	40.27	54.24	43.16	45.89	July 18	87.00	July 18	98	Dec. 24	-9.33	Dec. 24	-15	113
Beverly, N. Y	45.10	53.30	47.60	48.67	June 30	83.00	June 30	89	Feb. 6	-4.67	Feb. 7	-12	101
Bladensburg, Md. (1)	46.35	62.19	50.43	52.99	July 19	86.17	July 19	102	Feb. 7	7.50	Feb. 11	3	99
Bloomfield, N. J	40.57	50.70	42.55	44.61	July 19	86.50	June 30	97	Feb. 6	-2.67	Feb. 6	-9	106
Brandon, Vt	40.57	50.70	42.55	44.61	June 30	83.16	June 30	90	Feb. 6	-17.83	Feb. 7	-22	112
Burlington, Vt	41.49	52.10	43.47	45.69	June 30	84.60	{ June 30, July 17 }	92	Feb. 6	-19.00	Feb. 6	-24	116
Burlington, N. J	47.90	58.61	51.05	52.52	July 19	87.33	June 29	93	Feb. 7	3.67	Feb. 7	-2	95
Camden, S. C	56.24	71.53	60.60	62.79	April 19	86.67	April 19	97	Jan. 27, 30, 31	29.67	Feb. 1	16	81
Canton, N. Y	39.89	51.26	42.70	44.62	June 30	81.67	Aug. 2	93	Feb. 6	-27.00	Feb. 7	-40	133
Carmel, Me	37.06	51.28	37.61	41.98	July 2	80.00	July 1	96	Feb. 7	-12.67	Feb. 7	-24	120
Cedar Keys, Fa	66.60	73.21	69.42	69.74	Aug. 9	85.00	{ July 16, Aug. 12 }	91	Feb. 28	40.00	Feb. 27	32	69
Chapel Hill, N. C	53.01	68.64	57.65	59.77	{ June 30, July 19 }	86.67	April 19	102	Feb. 28	25.00	Feb. 27, 28	18	84
Charleston, S. C	61.17	71.96	64.80	65.98	....	....	Aug.	94	....	....	....	28	66
Cincinnati, Ohio	....	....	....	54.12	July 20	86.50	July 20	95	Dec. 26	7.00	Dec. 26	2	93

(1) Year complete, except October.

NAME OF STATION.	MEAN.				MAXIMA.				MINIMA.				RANGE FOR THE YEAR.
	7 A. M.	2 P. M.	9 P. M.	MEAN.	WARMEST DAY.		HIGHEST DEGREE.		COLDEST DAY.		LOWEST DEGREE.		
					Date.	Mean temp.	Date.	Temp	Date.	Mean temp.	Date.	Temp.	
Columbus, Miss	......	......	......	63.96	......	......	May	94½	......	......	Feb.	14½	80
Concord, N. H.	40.87	52.28	43.91	45.69	July 1	84.33	June 30, July 1	93	Feb. 6	-11.33	Feb. 7	-24	117
Cooper, Mich. (1)	46.67	55.15	44.58	48.80	June 29	82.67	May 22	92	Feb. 6	7.03	Feb. 6, 7	00	92
Detroit, Mich	44.56	53.50	45.86	47.97	July 16	84.33	June 28	92	Feb. 6	-7.00	Feb. 6	-14	106
Dubuque, Iowa	43.27	54.69	46.66	48.21	July 17, Aug. 3	86.33	May 22	99	Jan. 13	-2.67	Dec. 23, 24	-13	112
Easton, Pa	44.89	60.25	47.90	51.01	June 30	86.67	June 30	102	Feb. 6	-5.00	Feb. 6	-13	115
Exeter, N. H.	40.01	51.26	41.48	44.25	July 17	79.67	July 17	90	Feb. 6	-11.33	Feb. 7	-20	110
Flint, Mich	41.02	54.79	43.07	46.29	......	......	June 29	93	......	......	Feb. 7	-31	124
Fort Madison, Iowa	44.48	58.21	50.61	51.10	July 18	91.33	July 19	102	Dec. 23	-7.00	Dec. 30	-18	84
Frederick, Md	48.81	59.55	51.53	53.30	June 29	89.00	June 30	95	Feb. 7	5.43	Feb. 7	4½	90½
Germantown, Ohio	46.34	57.88	48.19	50.80	July 19	84.58	July 19	93½	Dec. 26	-2.20	Dec. 26	-7	100½
Gettysburg, Pa	46.32	57.82	49.43	51.19	June 29	86.00	June 29, 30, July 19	94	Feb. 7	1.33	Feb. 6	-5	99
Glenwood, Tenn.	52.16	64.28	55.84	57.43	May 22	82.80	May 22	92½	Dec. 26	12.87	Dec. 26	3	89½
Gouverneur, N. Y	41.09	51.76	44.27	45.71	July 1	83.30	July 1, Aug. 3	90	Feb. 5	-26.00	Feb. 6, 7	-44	134
Grand Rapids, Mich	40.45	53.12	44.52	46.03	July 16	84.67	July 16	93	Feb. 6	0.67	Feb. 6	-8	101
Granville, Ohio	46.65	57.75	49.28	51.23	July 18	87.67	June 28, 29, July 15	91	Feb. 26	3.67	Dec. 26	1	90
Harrisburg, Pa	49.24	57.75	53.50	53.50	June 29, 30	89.33	June 30	97	Feb. 7	1.33	Feb. 7	-1	98
Hillsborough, Ohio	46.39	57.47	49.43	51.10	July 18	81.00	July 18, 19	90	Feb. 26	2.50	Dec. 26	-3	93
Jacksonville, Fa	65.66	75.46	66.59	69.24	May 23	91.00	May 23	101	Jan. 31	37.00	Feb. 1	30	71
Key West, Fa	72.84	79.02	......	75.93	June 8	86.50	June 8	93	Mar. 2	52.00	Jan. 23	52	41
Knox Hill, Fa. (1)	62.49	75.30	65.30	67.70	Aug. 15	86.00	May 24	98	Feb. 26	32.33	Feb. 27	25	......
Lewisburg, Va	48.23	62.45	50.98	53.89	July 19	79.00	May 23, April 18	92	Feb. 27	17.00	Feb. 27	12	80
Lima, Pa	47.38	57.40	48.95	51.24	June 30	86.20	July 19	93	Feb. 6	3.63	Feb. 7	-1	94
Lodi, N. Y	42.46	51.26	43.23	45.65	June 30, July 19	83.00	July 19	90	Feb. 6	-18.33	Feb. 6	-22	112

(1) Year complete, except December.

NAME OF STATION.	MEAN.				MAXIMA.				MINIMA.				RANGE FOR THE YEAR.
	7 A. M.	2 P. M.	9 P. M.	MEAN.	WARMEST DAY.		HIGHEST DEGREE.		COLDEST DAY.		LOWEST DEGREE.		
					Date.	Mean temp.	Date.	Temp.	Date.	Mean temp.	Date.	Temp.	°
Manchester, Ill	46.16	59.23	49.04	51.48	June 28	85.83	July 18	96	Feb. 25	3.67	Dec. 26	–17	113
Manchester, N. H	40.45	54.78	46.18	47.14	July 1	87.00	{ June 30 { July 26	96	Feb. 6	–15.33	Feb. 7	–22	118
Mendon, Mass	43.34	52.58	45.45	47.12	July 1	87.00	{ June 30 { July 19	92	Feb. 6	–10.33	Feb. 7	–17	109
Milton, Ind	47.53	56.84	49.27	51.21	July 18	85.33	July 18	94	Dec. 26	0.66	Dec. 26	–8	102
Milwaukie, Wis., (Winkler)	40.06	50.21	40.42	43.56	July 18	84.00	July 18	94	Dec. 24	–8.67	Dec. 24	–14	108
Morrisville, Pa	45.77	57.64	48.18	50.53	July 19	83.67	July 19	94	Feb. 6	3.33	Feb. 7	–4	98
Muscatine, Iowa	42.33	56.47	43.85	47.55	July 17	84.67	Aug. 3	96	Dec. 23	–9.00	Jan. 23	–23	119
Nantucket, Mass	49.65	53.93	48.41	50.66	July 18	78.33	July 18	86	Feb. 6	8.33	Feb. 6	4	82
Newark, N. J	....	....	....	50.68	July 19	85.63	June 29	96½	Feb. 7	2.12	Feb. 7	–8	104½
New Bedford, Mass	45.07	53.97	46.12	47.18	June 29	80.67	July 1	89	Feb. 6	–7.83	Feb. 6	–14	103
Newburyport, Mass	43.44	51.36	45.11	46.64	July 17	82.90	July 19	94	Feb. 6	–6.00	Feb. 7	–13	107
New Harmony, Ind	50.80	62.51	53.56	55.62	July 19	87.67	July	95	Dec. 26	8.00	Dec. 26	–2	97
New Orleans, La	62.66	72.80	66.72	67.39	Aug. 13	84.17	May 28, 29	93½	Feb. 26	34.67	Dec. 26	26	67½
New Wied, Texas	62.20	79.50	64.84	68.85	Aug. 7	88.67	May 19	103	Dec. 24	23.67	Dec. 26	13	90
Norristown, Pa	46.57	56.60	48.40	50.52	July 19	83.67	June 29	92	Feb. 6	3.33	Feb. 7	–2	94
North Attleboro, Mass	40.65	53.79	43.63	46.02	July 19	84.33	July 19	93	Feb. 6	–9.00	Feb. 7	–15	108
Oberlin, Ohio	46.01	56.38	46.47	49.62	June 29	84.67	July 17	94	Feb. 6	2.33	Feb. 24	–10	104
Ottawa, Ill	44.15	56.84	45.83	48.94	June 29	88.33	May 22	96	Dec. 24	–3.00	Dec. 30	–10	106
Oxford, Miss	57.27	68.33	59.43	61.68	May 22	86.33	Aug. 13	94	Dec. 25	14.67	Dec. 26	8	86
Pella, Iowa	39.82	56.36	42.95	46.04	July 17	89.00	{ July 18 { Aug. 3	98	Dec. 24	–11.33	Dec. 29	–16	114
Penn Yan, N. Y	39.90	52.63	45.67	46.07	July 19	81.33	{ June 30 { July 18, 19	90	Feb. 6	–14.00	Feb. 7	–21	111
Perry, Me	36.75	46.89	39.91	41.18	July 19	71.33	July 19	82	Feb. 6	–8.00	Feb. 7	–15	97
Perrysburg, Ohio	48.72	57.96	49.57	52.08	June 29	88.00	July 16	94	Feb. 6	3.67	Feb. 6	–7	101
Philadelphia, Pa	50.42	58.68	53.80	54.30	June 30	88.67	July 19	95	Feb. 6	4.50	Feb. 7	–1	96
Pittsburg, Pa., (Oak. Sta.)	45.45	55.35	47.76	49.52	June 30	82.67	{ June 29 { July 19	91	Feb. 6	3.33	Feb. 6	–9	100
Do....(Marine Hospital)	48.02	55.60	52.35	51.99	June 29	84.00	{ June 30 { July 19	90	{ Feb. 27 { Dec. 27	13.00	{ Feb. 27 { Dec. 27	10	80

NAME OF STATION.	MEAN.				MAXIMA.				MINIMA.				RANGE FOR THE YEAR.
					WARMEST DAY.		HIGHEST DEGREE.		COLDEST DAY.		LOWEST DEGREE.		
	7 A. M.	2 P. M.	9 P. M.	MEAN.	Date.	Mean temp.	Date.	Temp.	Date.	Mean temp.	Date.	Temp	
Platteville, Wis.	41.73	56.78	45.94	48.15	July 18	88.67	May 22	98	Dec. 23	–14.67	Dec. 24	–22	120
Pocopson, Pa.	49.54	59.35	50.06	52.98	June 30	87.00	July 19	98	Feb. 6, 7	4.33	Feb. 7	–2	100
Pomfret, Conn.	41.17	51.07	44.47	45.57	July 19	82.00	{June 30 July 19}	88	Feb. 6	–9.00	Feb. 7	–16	104
Pottsville, Pa.	44.65	55.93	45.26	48.61	July 17	88.00	July 17	98	Feb. 6	–4.00	Feb. 7	–6	104
Poultney, Iowa	40.09	54.80	41.47	45.45	July 16	85.67	July 12	100	Feb. 25	–7.33	Dec. 30	–18	118
Princeton, Mass.	40.54	48.97	42.20	43.90	July 1	79.00	July 2	91	Feb. 7	–5.57	Feb. 7	–17	108
Sag Harbor, N. Y. (1)	47.37	59.40	48.35	51.71	{June 30 July 19}	85.00	July 19	96	Feb. 6	–0.33	Feb. 6	–6	102
St. James, Mich.	40.35	48.98	39.71	43.01	June 30	82.33	June 30	90	Feb. 5	–9.83	Feb. 5	–16	106
St. Martins, Canada East.	36.29	50.04	39.78	42.04	June 30	84.76	June 30	94	Feb. 6	–25.23	Feb. 6	–34	128
Savannah, Ga.	60.47	73.78	64.46	66.24	Aug. 11	87.17	April 19	99½	Feb. 27	32.73	Jan. 27	27	72½
Schellman Hall, Md.	47.86	59.87	49.75	52.49	June 29	87.67	{June 29 July 20}	91	Feb. 7	2.33	Feb. 7	–2	93
Smithfield, Va.	....	....	....	57.46	June 29	85.75	July 20	96	Feb. 10	16.50	Feb. 10	10	86
Smithsonian Institution.	50.06	61.08	53.51	54.88	June 29	86.67	April 25	97	Feb. 7	6.50	Feb. 7	4	93
Smithville, N. Y.	39.85	50.76	42.33	44.31	July 17	79.33	July 1	94	Feb. 6	–26.67	Feb. 6	–38	122
Sparta, Ga.	54.25	72.55	56.02	60.94	May 23	86.33	May 23	100	Feb. 27	26.00	Feb. 27	18	82
Spencertown, N. Y.	41.92	50.94	43.22	45.36	July 1	84.67	June 30	95	Feb. 6	–13.33	Feb. 7	–19	114
Springfield, Mass. (2)	42.89	55.36	46.45	48.23	June 30	86.33	July 19	93	Feb. 6	–6.00	Feb. 7	–19	112
Steuben, Me.	39.62	46.98	40.85	42.48	June 29	77.00	June 29	87	Feb. 6	–9.33	Feb. 7	–19	106
The Rock, Ga..	....	....	....	60.53	Aug. 15, 16	81.00	....	....	Jan. 30	23.00	....	....	....
Upper Alton, Ill.	47.54	60.11	51.10	52.92	July 18	83.00	July 19	93	Dec. 24	0.33	Dec. 30	–8	101
Wampsville, N. Y.	42.32	54.43	43.53	46.76	July 29	84.00	June 29	94	Feb. 6	21.00	Feb. 6	–21	115
Warrington, Fa. (2)	64.03	72.08	70.16	68.76	Aug. 27	85.00	July 10	88	Dec. 26	33.33	Dec. 26	23	85
West Haverford, Pa.	46.76	58.96	....	51.91	June 30	87.00	July 30	95	Feb. 7	–2.00	Feb. 7	–4	99
Whitemarsh Island, Ga.	61.20	71.49	63.92	65.54	Aug. 11	86.00	April 20	93	Feb. 27	32.33	Jan. 27	25	68
Williamstown, Mass. (2)	40.21	50.23	43.44	44.63	June 30	81.50	June 30	87	Feb. 6	–14.17	Feb. 7	–18	107
Worcester, Mass.	42.80	52.71	44.81	46.77	June 30	85.33	July 1	91½	Feb. 6	–10.33	Feb. 7	–16½	108

(1) Year complete, except November. (2) For hours of observation, see monthly tables.

NAME OF STATION.	Mean Force of Vapor.			Relative Humidity.								
				MEAN.			MAXIMA.			MINIMA.		
	7 A. M.	2 P. M.	9 P. M.	7 A. M.	2 P. M.	9 P. M.	7 A. M.	2 P. M.	9 P. M.	7 A. M.	2 P. M.	9 P. M.
Aiken, S. C.	.453	.467	.461	81	55	71	100	100	100	34	00	19
Alexandria, Va.	.381	.440	.403	87	72	83	100	100	100	26	18	00
All Saints, S. C.	.495	.534	.519	84	67	81	100	100	100	45	16	17
Ann Arbor, Mich.: (Winchell)	.281	.308	.288	83	67	83	100	100	100	28	18	32
Amherst, Mass.	.301	.322	.306	90	67	85	100	100	100	50	22	39
Auburn, Ala.	.389	.367	.383	70	45	61	100	100	100	10	00	15
Austin, Texas	.460	.429	.485	86	47	76	100	100	100	43	00	15
Bloomfield, N. J.	.307	.347	.331	90	62	80	100	97	100	12	17	17
Burlington, Vt.	.248	.288	.252	71	57	77	100	100	100	00	00	00
Camden, S. C.	.438	.465	.458	81	55	75	100	100	95	13	10	30
Charleston, S. C.	....	....	....	.....	.....	.....	.....	.....	.....	.....	.....	.....
Deaf and Dumb Institution, N. Y.	.303	.324	.313	73	64	72	100	100	100	15	22	10
Dubuque, Iowa	.289	.309	.308	78	59	74	100	100	100	28	11	14
Frederick, Md.	.331	.367	.354	79	60	76	100	100	100	23	00	00
Gallipolis, Ohio	.393	.465	.434	82	68	79	100	100	100	43	00	42
Germantown, Ohio	.338	.360	.347	86	62	84	100	100	100	26	18	35
Glenwood, Tenn.	.394	.412	.448	84	57	78	100	100	100	24	14	25
Hillsborough, Ohio	.353	.394	.361	92	71	86	100	100	100	44	04	31
Jacksonville, Fa.	.606	.711	.626	88	77	90	100	100	100	00	39	47
Lima, Penn.	.315	.322	.327	82	61	81	100	100	100	31	10	34
Manchester, Ill.	.354	.464	.378	89	75	89	100	100	100	06	00	42
Nantucket, Mass.	.301	.295	.308	73	61	80	100	100	100	00	00	20
New Bedford, Mass.	.313	.347	.325	86	71	87	100	100	100	50	26	00
Newburyport, Mass.	.267	.284	.282	76	59	76	100	100	100	04	04	04
New Harmony, Ind.	.397	.440	.429	88	66	87	100	100	100	42	13	39
New Orleans, La.	.560	.551	.583	88	63	82	100	100	100	10	16	27
New Wied, Texas	.550	.709	.587	82	64	86	100	100	100	00	00	00
Norristown, Pa.	.306	.338	.317	77	63	73	100	100	100	00	02	00
North Attleboro, Mass	.267	.274	.282	75	55	77	100	100	100	00	02	17
Oxford, Miss.	.464	.513	.477	84	67	81	100	100	100	12	07	18
Penn Yan, N. Y.	....	.285	....	.....	64	.....	.....	100	.....	.....	14	.....
Philadelphia, Pa.	.338	.396	.360	77	60	72	100	100	100	41	16	34
Pomfret, Conn.	.302	.358	.309	90	77	88	100	100	100	45	41	48
Princeton, Mass.	.284	.254	.246	77	63	74	100	100	100	00	09	00
St. Louis, Mo.	.350	.370	.372	77	56	74	100	97	100	00	09	20
St. Martins, Canada East	.240	.319	.263	86	73	84	100	100	100	22	28	26
St. Martins, C. E. (1)	.252	.336	.275	87	74	86	100	100	100	29	35	59
Savannah, Ga.	.504	.506	.520	80	54	75	98	97	96	37	11	24
Smithsonian Inst.	.358	.371	.387	82	59	79	100	100	100	24	00	34
Upper Alton, Ill. (2)	.399	.568	.437	87	84	84	100	100	100	02	31	25
Westfield, Mass.	.258	.283	.283	79	59	78	100	100	100	00	12	04

(1) The values are derived from the same observations as the preceding, by using the reductions of the observer, which were made by means of a different set of tables.

(2) Not fully reliable, from fewness of observations in the colder months.

NAME OF STATION.	7 A. M.	2 P. M.	9 P. M.	NAME OF STATION.	7 A. M.	2 P. M.	9 P. M.
Alexandria, Va.	5.52	5.40	4.33	Lodi, N. Y.	6.70	6.80	5.77
All Saints, S. C.	3.96	4.21	3.47	Manchester, Ill. (3)	5.16	5.52	4.27
Amherst, Mass.	5.31	5.42	4.66	Manchester, N. H.	5.64	6.02	5.13
Angelica, N. Y.	5.56	4.84	4.52	Mendon, Mass.	5.17	5.57	4.69
Ann Arbor, Michigan,				Milton, Ind.	4.75	4.57	3.92
(Woodruff)	5.91	5.69	4.94	Milwaukie, Wis.	4.77	4.43	3.75
Athens, Ill.	4.92	5.02	4.38	Morrisville, Pa.	4.12	3.79	3.98
Augusta, Ill.	4.91	4.43	3.83	Muscatine, Iowa	4.22	4.05	3.62
Austin, Texas	6.58	5.94	3.27	Nantucket, Mass.	5.00	4.65	4.72
Baldwinsville, N. Y.	6.15	6.40	5.29	New Bedford, Mass.	6.15	6.32	6.12
Battle Creek, Mich.	5.20	5.21	4.55	Newburyport, Mass.	7.25	6.21	7.03
Beloit, Wis.	3.48	2.89	2.87	New Harmony, Ind.	5.16	5.47	5.54
Beverly, N. Y.	5.61	5.79	4.86	New Orleans, La.	5.06	5.78	2.72
Bladensburg, Md. (1)	5.72	4.44	4.60	New Wied, Texas	3.95	5.90	7.14
Bloomfield, N. J.	4.74	4.86	4.25	Norristown, Pa.	5.37	5.84	4.74
Brandon, Vt.	5.84	5.12	4.85	North Attleboro, Mass.	4.98	5.07	4.26
Burlington, Vt.	3.39	3.07	3.12	Oberlin, Ohio	5.12	4.54	4.06
Burlington, N. J.	7.21	5.77	6.56	Ottawa, Ill.	3.90	4.16	4.01
Camden, S. C.	5.57	5.52	5.00	Oxford, Miss.	4.10	4.62	3.22
Canton, N. Y.	6.23	6.11	5.35	Penn Yan, N. Y. (3)	6.94	6.47	6.47
Carmel, Me.	5.22	5.53	4.62	Perry, Me. (3)	5.59	5.12	5.48
Cedar Keys, Fa.	3.87	4.22	2.88	Perrysburg, Ohio (3)	....	....	....
Chapel Hill, N. C.	4.62	4.07	5.27	Philadelphia, Pa.	6.51	6.45	5.10
Concord, N. H.	4.52	4.43	3.88	Pocopson, Pa.	6.32	6.49	5.13
Cooper, Mich. (2)	4.95	4.66	4.31	Pomfret, Conn.	5.32	5.19	4.93
Deaf and Dumb Insti-				Pottsville, Pa.	6.17	6.12	5.23
tution, N. Y.	5.53	5.64	4.72	Poultney, Iowa	5.52	5.57	4.88
Detroit, Mich.	6.52	6.62	5.14	Princeton, Mass.	5.86	6.14	5.01
Dubuque, Iowa	5.47	4.86	3.98	Sag Harbor, N. Y. (4)	5.06	5.16	4.54
Easton, Pa.	5.12	5.81	4.34	St. James, Mich.	6.12	5.54	5.67
Exeter, N. H.	4.85	4.81	4.17	St. Louis, Mo.	5.30	4.66	4.07
Fort Madison, Iowa (3)	5.18	4.67	4.52	St. Martins, Canada E.	4.36	4.74	4.72
Frederick city, Md.	5.93	5.60	4.96	Savannah, Ga.	4.72	5.11	3.48
Fryeburg, Me.	4.90	4.60	4.73	Schellman Hall, Md.	5.02	5.02	5.11
Germantown, Ohio	6.42	6.53	5.02	Smithville, N. Y.	5.72	5.89	5.27
Gettysburg, Pa.	5.46	5.42	4.76	Southwick, Mass.	5.42	5.99	6.29
Glenwood, Tenn.	5.77	6.26	3.69	Sparta, Ga.	4.82	5.60	3.86
Gouverneur, N. Y.	6.27	6.77	5.56	Spencertown, N. Y.	4.83	4.88	4.66
Grand Rapids, Mich.	5.35	5.23	3.98	Springfield, Mass (3)	4.78	4.85	4.18
Granville, Ohio	5.58	5.86	4.88	Steuben, Me.	5.98	6.28	6.27
Harrisburg, Pa.	5.72	5.47	5.12	Upper Alton, Ill.	3.73	3.75	2.27
Hillsborough, Ohio	4.93	4.49	....	Wampsville, N. Y.	6.65	6.60	5.82
Jacksonville, Fa.	4.04	4.30	2.36	Warrington, Fa. (3)	3.93	3.91	3.70
Key West, Fa.	4.98	4.82	....	West Haverford, Pa.	4.78	6.71	....
Knox Hill, Fa. (2)	5.65	5.84	4.38	Whitemarsh Island, Ga.	3.84	4.48	3.37
Lewisburg, Va.	6.05	6.12	5.16	Williamstown, Mass. (3)	6.50	6.29	5.72
Lima, Pa.	5.67	6.03	4.91	Worcester, Mass.	4.47	4.62	3.99

(1) October wanting.
(2) December wanting.
(3) For hours of observation, see monthly tables.
(4) November wanting.

www.ingramcontent.com/pod-product-compliance
Lightning Source LLC
LaVergne TN
LVHW021426110826
845150LV00007B/2096

* 9 7 8 1 4 2 5 5 0 8 1 4 2 *